Progress in Molecular and Subcellular Biology

Series Editors: Ph. Jeanteur, Y. Kuchino, W.E.G. Müller (Managing Editor), P.L. Paine

19

Springer

Berlin
Heidelberg
New York
Barcelona
Budapest
Hong Kong
London
Milan
Paris
Santa Clara
Singapore
Tokyo

Werner E.G. Müller (Ed.)

Molecular Evolution: Evidence for Monophyly of Metazoa

With 54 Figures

Springer

Professor Dr. WERNER E.G. MÜLLER

Institut für Physiologische Chemie
Abteilung für Angewandte Molekularbiologie
Johannes Gutenberg-Universität
Duesbergweg 6
D-55099 Mainz
Germany

ISBN-13: 978-3-642-48747-7 e-ISBN-13: 978-3-642-48745-3
DOI: 10.1007/978-3-642-48745-3

Library of Congress Cataloging-in-Publication Data.
Müller, Werner E. G.
Molecular evolution : evidence for monophyly of metazoa/Werner E.G. Müller.
p. cm.—(Progress in molecular and subcellular biology; 19)
Includes bibliographical references and index.
ISBN-13: 978-3-642-48747-7
1. Metazoa. 2. Phylogeny. 3. Evolutionary genetics.
4. Molecular evolution. I. Title. II. Series.
QH503.P76 no. 19QL45.2
572.8 s—dc21
[572.8'38]

Cover design: Meta Design, Berlin
Typesetting: Best-set Typesetter Ltd., Hong Kong
SPIN 10540565 39/3137 5 4 3 2 1 0 – Printed on acid-free paper

Preface

This volume focuses on the origin of multicellular animals, Metazoa. Fossil records are scarce and morphological, biochemical or cytological evidence has not demonstrated an unequivocal phylogeny. Therefore, the question remains: Did Metazoa evolve from Protozoa only once, or several times? Is the origin of animals monophyletic or polyphyletic? Especially the relationships between the existing lower metazoan phyla, particularly the Porifera, are uncertain. In addition, phylogenetic analyses of nucleotide sequences of small and large ribosomal RNA also revealed no conclusive results with regard to the evolutionary dichotomy of diploblasts (Porifera, Cnidaria and Ctenophora) and triploblasts in general and the diploblastic lineages in particular.

During the evolution of unicellular eukaryotes to multicellular animals, complex molecular systems were established which allow a tuned and controlled interaction between cells among each other, and cells and extracellular molecules. Hence, cell surface-associated adhesion and growth hormone receptors, as well as their corresponding extracellular ligands, have evolved. In addition, transcription factors had to be introduced that control gene functions in the variety of cell types in developing and adult organisms.

Only in the last few years some groups have concentrated their research activities on the cloning of genes encoding such proteins which are typical for multicellularity, such as adhesion molecules and receptors as well as transcription factors, from the lowest metazoan phylum, the Porifera. Comparisons of the deduced amino acid sequences revealed an unexpectedly high conservation of the functional domains in the molecules from sponges up to vertebrates. These data support the monophyly of Metazoa, including the phylum Porifera.

Progress in the field of molecular evolution, especially with respect to the identification of genes encoding "metazoan proteins" in lower phyla, is rapid. This volume summarizes the status achieved until 1997. Hopefully, the exploration in this field of early "metazoan proteins" will continue and may thus also shed some light on the first existing metazoans which appeared approximately 800 million years ago.

Mainz, Germany
August 1997

W.E.G. Müller

Contents

Transition from Protozoa to Metazoa: An Experimental Approach

Werner E.G. Müller and Isabel M. Müller[1]

1 Introduction

Until recently, stromatolites were thought to be the oldest fossils on earth that were very abundant 2000 to 3000 Ma (million years) ago (Walter 1994). Recently, the biological origin of these fossils has been questioned (Walter 1996). The universal phylogenetic tree exhibits a tripartite division of the living world into Bacteria ("eubacterial"), Archaea ("archebacterial"), and Eucarya ["eukaryotic" (Woese 1987; Woese et al. 1991)]. Based on comparisons of amino acid (aa) sequence data from enzymes, it has been proposed that the common ancestor of prokaryotes and eukaryotes lived about 2000 Ma ago (Doolittle et al. 1996). Phylogenetic analysis of the 70 kDa heat-shock proteins suggested that the first eukaryotic cell was a fusion product of a member of the Archaebacteria with a Gram-negative bacterium. Later in evolution a purple proteobacterium and a cyanobacterium invaded eukaryotic cells and gave rise to mitochondria and chloroplasts (Gupta and Singh 1994).

The sponges (Porifera) are the oldest metazoan phylum; they appeared – based on aa sequence data – approximately 800 Ma ago (Müller et al. 1994; Müller 1995). The oldest isolated sponge spicules identified so far were found in thin-sectioned material from the Dengying Formation (Proterozoic: Sinian), "Shibantan" Member, Hubei Province (Steiner et al. 1993); these are triaenes and aster microscleres of the probably already modified steraster type (Reitner and Mehl 1995), a constituent character of the taxon Geodiidae, which is thus the oldest known group within the Demospongiae, 550 Ma ago.

It is fortunate that most molecular sequence data available from sponges until now are from *Geodia cydonium*, which belongs to the family of Geodiidae (Müller 1995; Mehl et al. 1997). Therefore, where possible, reference is made to sequences from this sponge species if comparisons with metazoan organisms are drawn.

[1] Institut für Physiologische Chemie, Johannes Gutenberg-Universität, Abteilung Angewandte Molekularbiologie, Duesbergweg 6, 55099 Mainz, Germany

Progress in Molecular and Subcellular Biology, Vol. 19
W.E.G. Müller (Ed.)

The period required for the evolution of Protozoa to Metazoa was more than 1000 million years. This comparatively long time is amazing considering the fact that the basic molecules establishing and maintaining life were already present in all three regna, Bacteria, Archaea, and Eucarya, which diverged approximately 3500 Ma ago (Knoll 1994).

2
Proteins Present in All Biotic Regna

Life is characterized by the ability to (1) reproduce; (2) grow and develop; (3) utilize energy; (4) respond to the environment; (5) maintain homeostasis; and (6) allow evolutionary adaptation. Therefore, it is not surprising that basic proteins required to maintain these abilities have been conserved in all three regna. Some examples of molecules with high homology between the regna are given here.

2.1
Reproduction

Following the assumption that RNA-protein-based cells precede DNA-RNA-protein based cells (Lazcano 1994), it is conceivable that the RNA polymerases in the three regna share similar characteristics (Zillig 1991).

On the level of protein synthesis, sequence alignments revealed that the elongation factor 1α (EF1α), which assists binding of activated tRNAs to ribosomes, is conserved throughout the three regna (Baldauf and Palmer 1993). As examples, the alignment of the aa sequences coding for the EF1α from the archaebacterium *Methanococcus vannielii*, the bacterium *Escherichia coli*, and the sponge *G. cydonium* – as a representative of eukaryotic animals – is shown in Fig. 1. The sponge sequence shares 57% identity with the archaebacterial and 31% with the bacterial sequence.

2.2
Growth and Development

Growth and development require, e.g., controlled gene expression mediated by distinct promoters; some promoters are known to be conserved among the regna (Reiter et al. 1990).

2.3
Energy Utilization

The ATP synthase, the central enzyme of energy conservation (Konishi et al. 1990), and enzymes, e.g., of glycolysis (Hensel et al. 1989), are likewise highly homologous among the three regna.

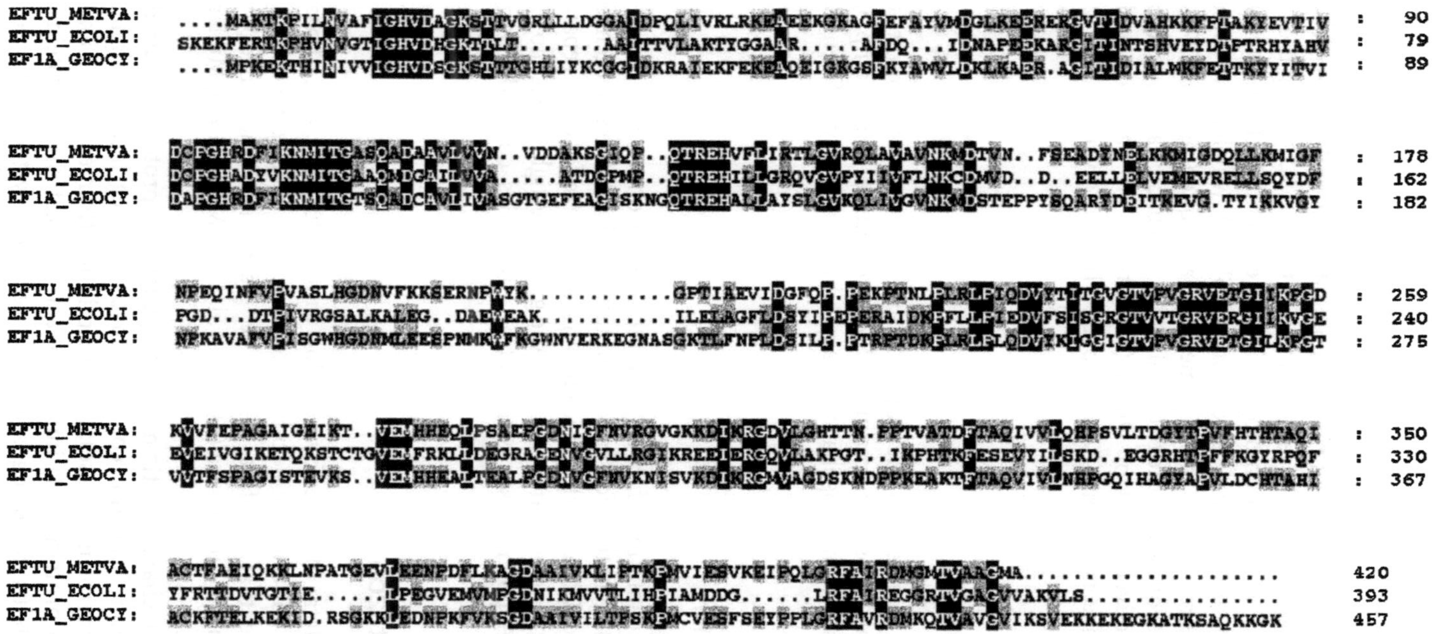

Fig. 1. Alignment of the deduced aa sequences of the elongation factor-1α (EF1α) from the archaebacterium *M. vannielii* [*EFTU_METVA*, P07810 (Lechner et al. 1988)], the bacterium *E. coli* [*EFTU_ECOLI*, P02990 (Jones et al. 1980)] as well as the sponge *G. cydonium* [*EF1A_GEOCY*, Y09023 (to be published)]. Identical residues in all three sequences are shown in *inverted type*, whereas residues conserved in half of the sequences are *shaded*

2.4
Response to the Environment

Organisms respond to organismic and environmental stress with the expression of heat shock proteins (HSP; survey in Yamane et al. 1995). Among them are two families, the HSP70s (the homologues in Bacteria and Archaebacteria are the DnaK proteins) and the DnaJ (also termed HSP40s), which share high homology among the three regna. The alignments of aa sequences of HSP70s (identity of the sponge sequence to the bacterial and archaebacterial sequences is 51 and 52%, respectively) and of DnaJs (identity to the bacterial and archaebacterial sequence is 28 and 27%, respectively) are given in Figs. 2 and 3.

2.5
Homeostasis

Homeostasis against changes in the internal and external environment is maintained, e.g., by allosteric enzymes. Some of these, for example, glyceraldehyde phosphate dehydrogenase, are again highly conserved (Hensel et al. 1989).

2.6
Evolutionary Adaptation

Prerequistites for the adaptation of an organism to its environment are stability as well as changes in genetic information (mutations). With the emergence of the double-stranded DNA world, mechanisms were introduced – repair systems – to detect and correct damage in DNA, e.g., by UV light (Lazcano et al. 1988). Hence, DNA polymerases with editing properties could develop (Lazcano 1994).

3
Monophyly of Metazoa

Based on sequence data of genes typical for multicellularity, e.g., those sequences coding for adhesion molecules/receptors and nuclear receptors, we could demonstrate that Porifera should be placed into the kingdom Animalia together with the Eumetazoa (Müller et al. 1994; Müller 1995; Müller 1997).

The structures of the characteristic metazoan genes and proteins required for (1) tissue formation (galectin, collagen, myosin, integrin); (2) signal transduction (receptor tyrosine kinase); (3) transcription (homeodomain and MADS-box-containing proteins); (4) immune reaction (heat shock proteins, proteasome, and proteins featuring scavenger receptor cysteine-rich domains); and (5) sensory organs (crystallin) have been identified in *G. cydonium*

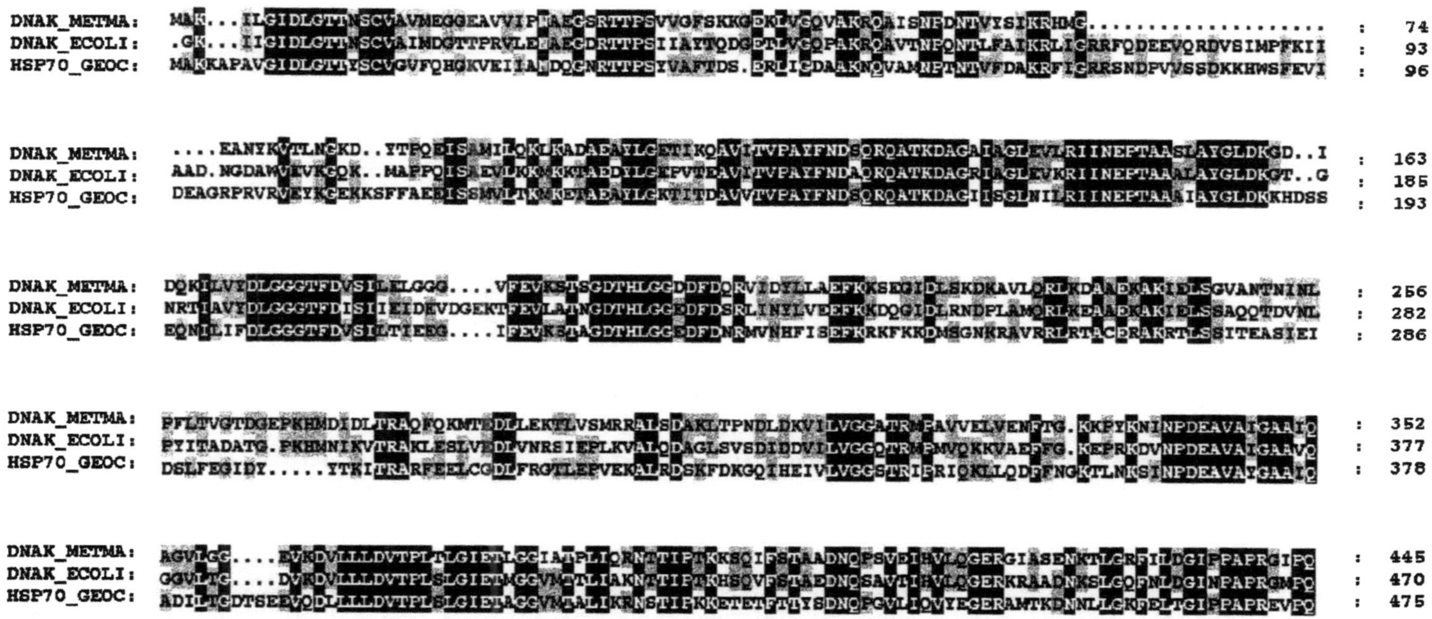

Fig. 2. Alignment of the deduced aa sequences of the heat shock protein 70 (HSP70) from the archaebacterium *Methanosarcina mazei* [*DNAK_METMA*, P27094 (Macario et al. 1991)], the bacterium *E. coli* [*DNAK_ECOLI*, P04475 (Bardwell and Craig 1984)] as well as the sponge *G. cydonium* [*HSP70_GEOC*, Y09037 (Koziol et al. 1996)]. Identical residues in all three sequences are shown in *inverted type*

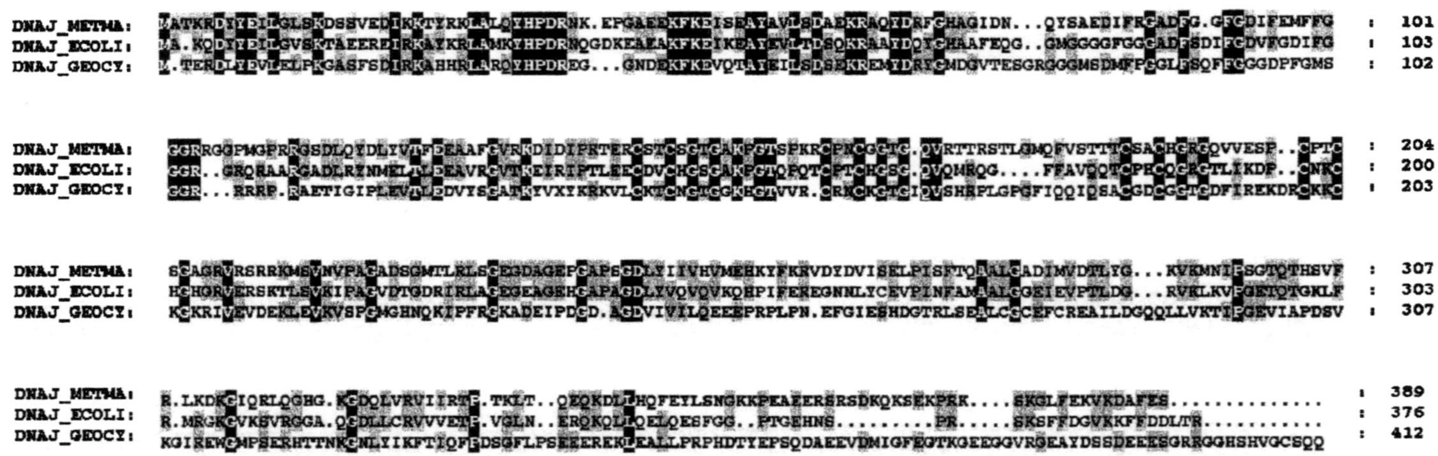

Fig. 3. Alignment of the deduced aa sequence for DnaJ from the archaebacterium *M. mazei* [*DNAJ_METMA*, P35515 (Macario et al. 1993)], the bacterium *E. coli* [*DNAJ_ECOLI*, P08622 (Ohki et al. 1986)] as well as the sponge *G. cydonium* [*DNAJ_GEOCY*, X94985 (Koziol et al., 1997)]. Identical residues in all three sequences are shown in *inverted type*

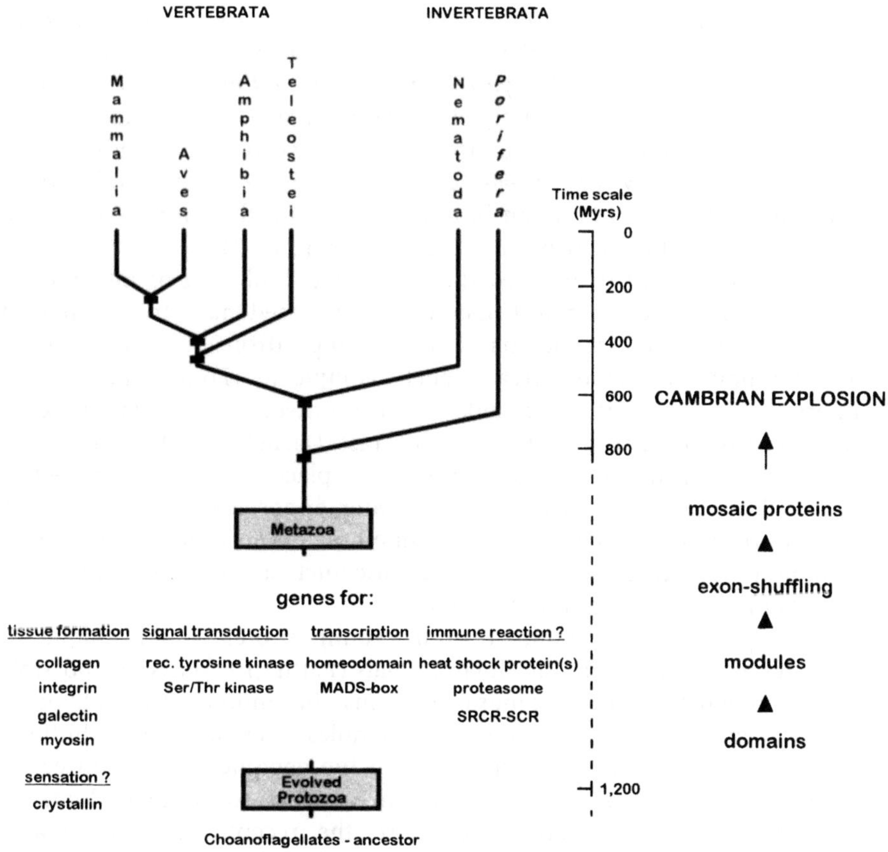

Fig. 4. Phylogenetic relationship of Porifera within the animal groups based on molecular biological data, obtained from sequences of "metazoan" proteins required for tissue formation, signal transduction, transcription, immune reaction (potential) and sensory organs (potential). It is proposed that the Cambrian explosion of metazoan radiation became possible after the creation of the evolutionary mechanism of modularization of distinct protein domains, thus allowing the formation of mosaic proteins by exon shuffling; this process happened approximately 1000 Ma ago. It is assumed that Metazoa originated from evolved Protozoa, e.g., Choanoflagellata (syn. Craspedomonadida; Protozoa)

(Fig. 4) and found to display high homology to sequences from members of higher metazoan phyla (Müller 1997).

4 Origin of Metazoan Genes and Proteins

4.1 Introns

Genes of eukaryotes are split into exons and introns (Berger et al. 1977; Williamson 1977). A correlation between intron density and developmental complexity in eukaryotes has been established (Mattick 1994) and is based on the experimental finding that the number of introns per kilo base of coding region increases from one in *Dictyostelium* and *Plasmodium* to three or four in plants and fungi to approximately six in vertebrates (Palmer and Logsdon 1991). Parallel to this, the intron size increases also (Cavalier-Smith 1985). Support for the theory that at least some introns have developed functions of their own (Mattick 1994) can be seen in the fact that introns exhibit a nonrandom nucleotide distribution (Gutierrez et al. 1993).

Since the discovery of introns, the origin and evolution of intron-exon structures remain under debate. Gilbert (1978) proposed that introns appeared late in evolution, are descendants of self-splicing introns, and are involved in the shuffling of exons to increase genetic complexity (Long et al. 1995). In support of this, Blake (1978) extended the exon theory by correlating exons with "domains" present in proteins. In contrast, Darnell and Doolittle (1986) assumed that introns were present in the first place and already existed in genes of the progenote. This theory was later opposed by Cavallier-Smith (1991), who assumes that, after the symbiotic event which gave rise to mitochondria and chloroplasts, the eukaryotic nucleus was invaded by group II introns originating from the symbionts.

Mosaic proteins are assumed to be young proteins and unique to eukaryotes, especially to multicellular animals (Patthy 1994). As one possible way to explain the evolution of mosaic proteins, the "modularization hypothesis" was formulated, which assumes that modules suitable for exon shuffling are created by the insertion of introns into the genome adjacent to exons (Patthy 1994). Members of mosaic proteins contain extracellular domains, e.g., receptor proteins, among them also the receptor protein kinases (Ziegler et al. 1993). Mosaic proteins have been identified and described from metazoan phyla higher than sponges (Porifera; summarized by Patthy 1994, 1995).

Introns have been grouped into phase 1 introns (between the first and second nucleotide of a given codon), phase 2 introns (second and third nucleotide), and phase 0 introns (between codons; Patthy 1987).

Recently, we have cloned the gene coding for the oldest metazoan mosaic protein, the receptor tyrosine kinase (PTK) from the marine sponge *G. cydonium* (Schäcke et al. 1994a,b; Müller and Schäcke 1996). Cloning of this gene revealed that it contains only two introns which are present in the extracellular part of the protein (Gamulin et al. 1997). All other RTK genes studied contain introns in their tyrosine kinase (TK) domain also. Some of the genes encoding RTKs are over 100 kb long with more than 20 introns, as in the case of insulin receptor subfamily of RTKs (Seino et al. 1989). Phylogenetic analysis of the sponge TK domain indicates that this enzyme branched off first from the common tree of metazoan TK proteins (Gamulin et al. 1997).

4.2 Protein Modules

Modules are defined, according to Patthy (1987), as DNA segments which are comprised of exons, which code for distinct protein domains. In addition, these domains are associated with exon shuffling and exon duplication. It has been suggested that the large number of protein families (Chothia 1992) was generated by combinations of a few protein modules (survey in Baron et al. 1991). These modules usually have between 40 and 100 aa and are characterized by a typical structure and function (survey in Baron et al. 1991). The main protein domains have been grouped into nine superfolds (Orengo et al. 1994); among them are the globin and immunoglobulin superfolds.

5 Evolution of Metazoan Genes

Based on sequence data from phyla higher than Cnidaria it was deduced that mosaic proteins are constructed after the modularization of protein domains by intron insertions and subsequent dispersal through exon shuffling (Patthy 1994). This very constructive view implies that the ancestor molecules of Metazoa were without introns.

Experimental evidence for this assumption was lacking until the gene coding for the receptor tyrosine kinase from the sponge *G. cydonium* was cloned (Gamulin et al. 1997). Therefore, this gene can be considered to be the prototype of a metazoan mosaic protein. Two prototypes of composed proteins will be discussed here, which possibly indicate the trend of evolution to mosaic proteins.

5.1 Sponge: Receptor Tyrosine Kinase

Receptor protein tyrosine kinases (RTK) have been described only in higher invertebrate species as well as in vertebrate species (reviewed in Geer et al. 1994).

We cloned the RTK from a member of the lowest invertebrate phylum, the marine sponge *G. cydonium* (Schäcke et al. 1994a,b; Müller and Schäcke 1996). The deduced aa sequence of the RTK cDNA is comprised at the N-terminus in the potential extracellular part of the RTK of one Pro-Ser-Thr-rich domain, which is followed by two immunoglobulin-like (Ig-like) domains (Schäcke et al. 1994a,b; Pancer et al., in press); the transmembrane domain is followed by the juxtamembrane region and the tyrosine kinase (TK) domain (Fig. 5).

Cloning of the gene coding for the sponge RTK revealed that it contains only two introns, which are both present in the extracellular part of the protein (Fig. 5A; Gamulin et al. 1997). All other RTK genes studied contain introns in their TK domain as well. Some of the genes encoding RTKs are over 100 kb long with

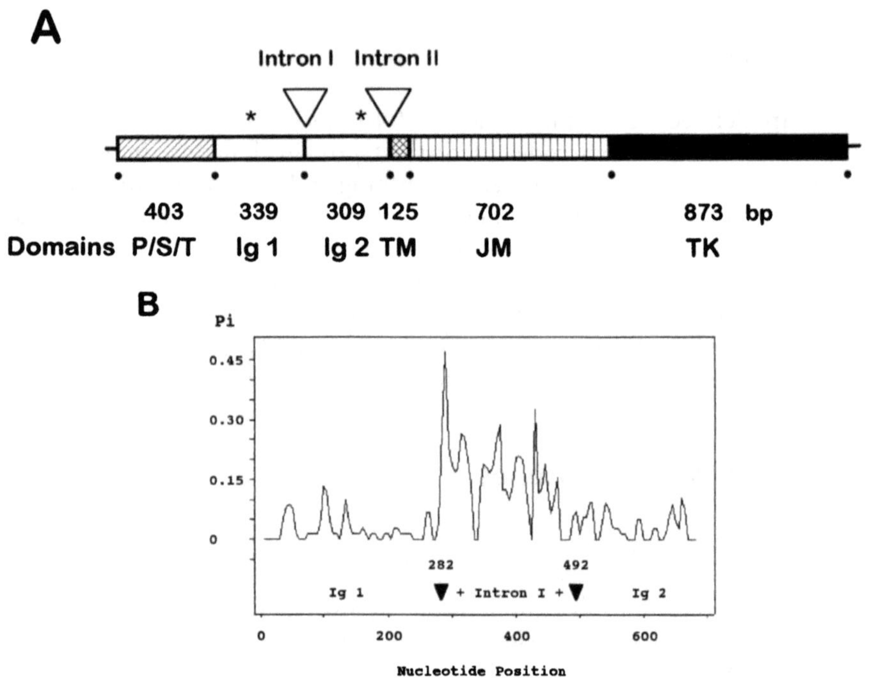

Fig. 5. A Schematic presentation of the RTK gene from *G. cydonium* (termed GCRTK). The coding region of the receptor consists of several domains: Pro-Ser-Thr rich domain (*P/S/T*), Ig-like 1 separated by an intron (*Intron I*) from Ig-like 2, then a second intron (*Intron II*), a transmembrane domain (*TM*), the juxtamembrane region (*JM*), and the TK domain (*TK*). Positions of the primers used for PCR amplification are indicated by *asterisks*. **B** Graphic presentation of averaged polymorphic nucleotide positions in the 13 independent clones from 4 different individuals. In the aligned 688-site-long sample, 131 polymorphic positions were analyzed (nucleotide diversity Pi = 0.06016); the graph was generated using window length 10 and step size 5 (DNA polymorphism, DnaSP; Rozas and Rozas 1995)

more than 20 introns, as in the case of the insulin receptor subfamily of RTKs (Seino et al. 1989).

To study possible intraspecies polymorphism in the two Ig-like domains of the RTK, we attempted genomic PCR amplification across the Ig-like domains, which are separated by an intron (intron I in Fig. 5A). Amplification products from four sponge specimens revealed several DNA bands ranging from 0.65 to 1.5 kb and notable differences in the intensity of the bands among individuals. To characterize some of these heterogeneous PCR products, 15 independent clones were analyzed from 4 individuals (Pancer et al., in prep.). Sequence analyses of the short insert clones revealed only two identical clones. Apart from these, 12 clones displayed pairwise nucleotide-sequence divergence of up to 23%. A graphic comparison of pairwise similarities among these sequences is presented in Fig. 5B. Further analysis of the distribution of polymorphic sites revealed that most of the sequence heterogeneity is confined to the region

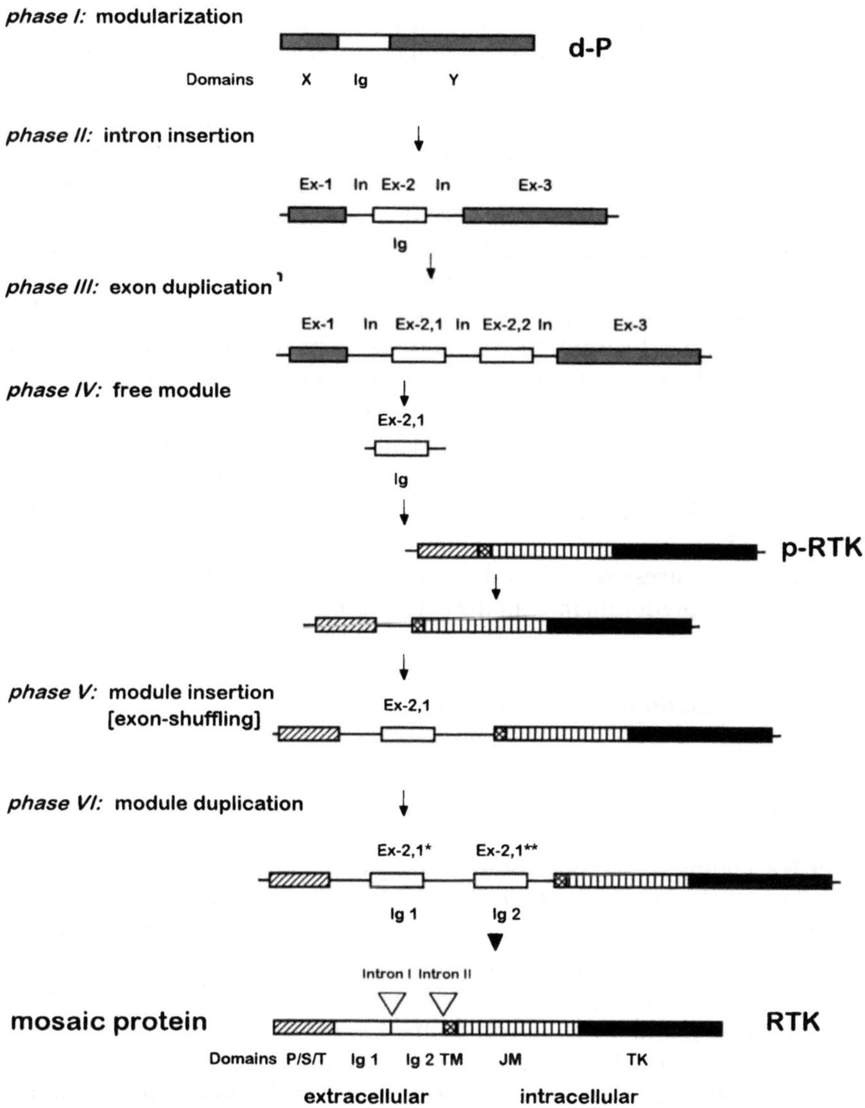

Fig. 6. Proposed generation of mosaic proteins; here, for example, the sponge RTK. The gene coding for a donor protein d-P which consists of an Ig-like domain (*phase I*) undergoes modularization (*phase II*) by insertion of introns (*In*). One of the created exons, Ex-2, is duplicated (*phase III*) and its product, Ex-2,1 is liberated (*phase IV*) and inserted as a module into the precursor gene of RTK (*p-RTK*); the process of module insertion by exon shuffling is completed (*phase V*). After duplication of the two modules, the sponge RTK gene, composed of two Ig-like domains (*Ex-2,1** and *Ex-2,1***), is completed (*phase VI*) and the mosaic protein formed (Modified after Patthy 1994)

within the boundaries of the intron (Fig. 5B), where some averaged nucleotide diversities reach a Pi of 0.47, nearly eight times the Pi of 0.06 for the whole sample.

The RTK gene of *G. cydonium* codes for a mosaic protein which is composed of several domains/modules: the Pro-Ser-Thr-rich domain, the two Ig-like domains, and a TK domain. The Ig-like domains are modules according to the strict nomenclature of Patthy (1995). Their coding exons are bordered by phase 1 introns (Gamulin et al. 1997). A schematic representation of the formation of the Ig-like module and its subsequent insertion into the RTK precursor gene are shown in Fig. 6. Modularization (phase I) of a postulated donor protein d-P occurs by intron insertion under formation of a protomodule "Ex-2" (phase II; Patthy 1994). In phase III, the protomodule undergoes duplication forming "Ex-2,1" and "Ex-2,2". Module "Ex-2,1" is excised in phase IV. In phase V, the excised "Ex-2,1" is inserted into the precursor gene RTK (p-RTK). Finally, module "Ex-2,1" is duplicated under formation of "Ex-2,1*" and "Ex-2,1**" (phase VI), creating the present existing gene, coding for the mosaic protein RTK.

It should be stressed that module or exon shuffling occurs, as predicted by Patthy (1995), predominantly in the extracellular portions of a mosaic protein.

5.2
Tunicate: Immunolectin – Antigen Receptor

5.2.1
Immunolectin

A cDNA coding for a mannose-binding lectin was cloned from a higher invertebrate, the tunicate *Botryllus schlosseri* (Ascidiacea); it is termed immunolectin (Pancer et al., 1997). The 333-aa-long sequence features three building blocks: (1) a Greek-key motif signature at the N-terminus; (2) a C-type lectin domain signature, consisting of the carbohydrate recognition domain (CRD); and (3) an Ig-like domain at the C-terminus. Ig-like domains were found as modules in a series of mosaic proteins (Williams and Barclay 1988); Hunkapiller et al. 1989; Schäcke et al. 1994b), but until now not in a lectin. In accordance with the formation of the RTK mosaic protein, we assume that the ancestral precursor of the immunolectin (p-ImLec) did not contain the Ig-like domain. The free module, comprised of the Ig-like domain, was probably excised from a donor molecule and inserted into the p-ImLec under formation of the present ImLec (Fig. 7A).

The *B. schlosseri* immunolectin is also interesting from another point of view. So far, it was believed that invertebrates are provided with an immune system which is optimized primarily to eliminate "nonself" challenges (Van de Vyver 1988). The major class of molecules involved in this reaction are the lectins, functioning with their carbohydrate recognition domains (Sima and Vetvicka 1993). In vertebrates this form of ancestral defense reaction is com-

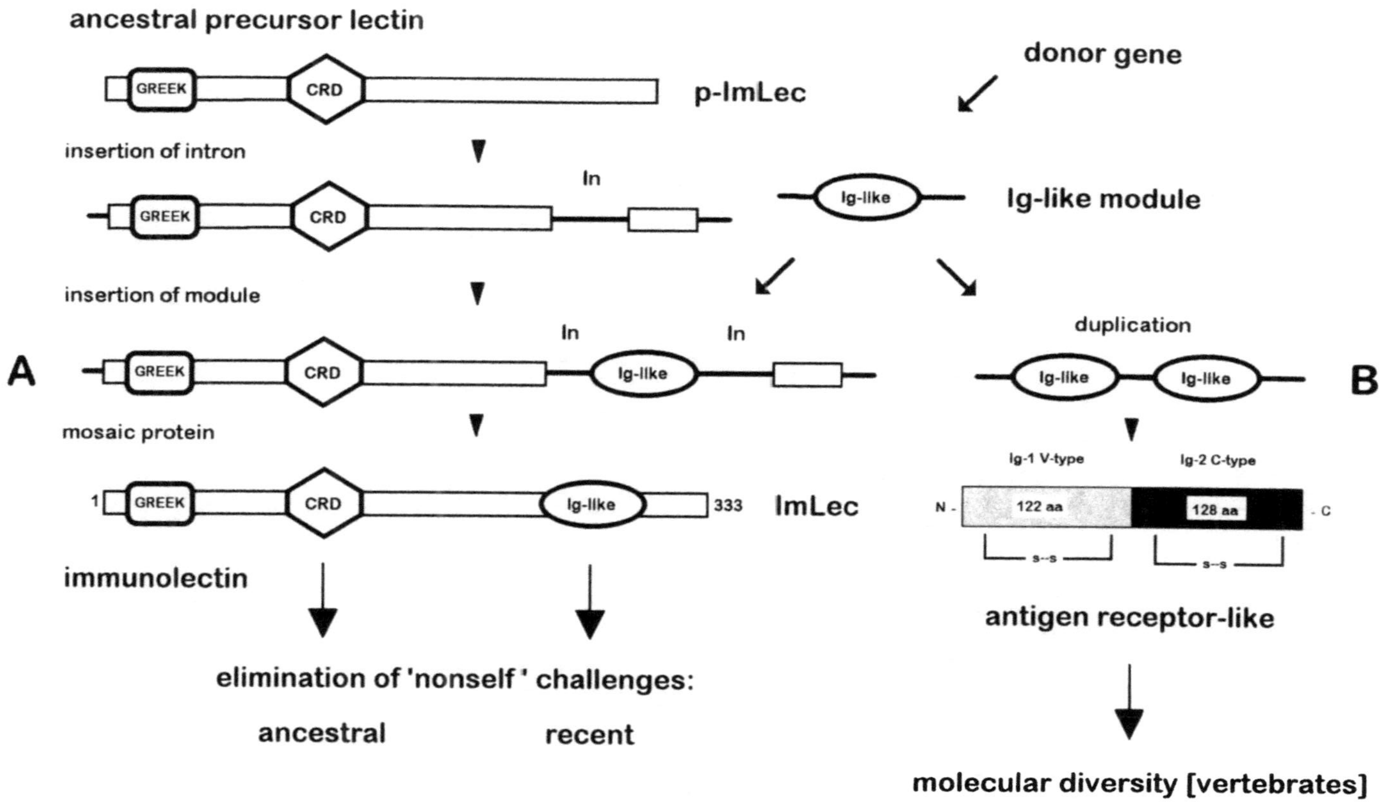

Fig. 7. **A** Proposed stages of generation of the mosaic immunolectin (*ImLec*) from the tunicate *B. schlosseri* by insertion of an Ig-like module into an intron-containing ancestral precursor *p-ImLec* gene. The deduced aa sequence of immunolectin consists of a Greek-key motif signature (*GREEK*) at the N-terminus, a C-type lectin domain signature with the carbohydrate recognition domain (*CRD*), and an Ig-like domain (*Ig-like*) at the C-terminus. The Ig-like domain is derived from a postulated second molecule, which underwent modularization, the donor gene. The Ig-like module was inserted into the p-ImLec gene after formation of an intron in the p-ImLec. It is postulated that the present ImLec protein combines features of the ancestral and of the recent type of elimination of "nonself" challenges. **B** As a "by-product" two free Ig-like modules are combined under formation of an antigen receptor-like gene. The latter molecule, also cloned from *B. schlosseri*, codes for both of a V- and a C-type Ig domain. This molecule could have given rise to immunoglobulins in vertebrates where it was subject to molecular diversification processes

plemented by effective adaptive immune systems, with immunoglobulins as key elements (Hildemann 1981). Hence, it can be speculated that the tunicate immunolectin represents the prototype of a molecule, functioning in an "ancestral" (via lectin binding) as well as in a "recent" (immunoglobulin-like) manner to eliminate nonself challenges (Fig. 7B).

5.2.2 *Antigen Receptor*

While invertebrates possess elaborate mechanisms to recognize self-antigens (Cooper et al. 1992; Smith and Davidson 1992; Sima and Vetvicka 1993; Humphreys and Reinherz 1994), it appears that the propensity for nonself discrimination by virtue of T-cell receptors and antibodies evolved later, perhaps after the divergence of jawless and jawed vertebrates (Marchalonis and Schluter 1990). As mentioned, molecules featuring Ig-like domains appeared early in eukaryotic evolution, e.g., yeast α-agglutinin cell wall-associated protein (Chen et al. 1995), sponge RTK (Müller 1995), and soluble proteins, such as the insect hemolin (Sun et al. 1990; Zhao and Kansot 1996) or the tunicate C-type lectin, with 14 of its 35 N-terminal residues identical to the variable region of Ig kappa (Suzuki et al. 1990).

Recently, a molecule from the tunicate *B. schlosseri* was cloned that features structural similarity to the vertebrate antigen receptors (Pancer et al. 1996b). The deduced aa sequence of the full-length cDNA, termed *B. schlosseri* "Possible Antigen Receptor-like Molecule", comprises two sets of β-strands (Fig. 7B) which are characteristic of Ig domains (Williams and Barclay 1988; Hunkapiller et al. 1989). The N-terminal domain (aa 16–138) appears to belong to the V-type, and the C-terminal (aa 139–267) appears to be a C-type of Ig-like domains.

Here, it is proposed that this novel "Possible Antigen Receptor-like Molecule" is a "by-product" of the generation of the mosaic protein, the immunolectin (Fig. 7B). It is thought that the antigen receptor-like molecule is generated by combining two free Ig-modules.

5.3 Protein Kinases

The largest known protein family includes protein kinases, which transfer ATP (or GTP) to proteins (Hardie and Hanks 1995). They are subdivided into those enzymes which transfer the phosphate group to the aa Ser and Thr, the Ser/Thr kinases, and those which transfer the Tyr kinases to Tyr as the acceptor group. Most protein kinases have been identified from eukaryotes (Hardie and Hanks 1995).

5.3.1 *Ser/Thr Kinases*

Two Ser/Thr kinases were cloned from the sponge *G. cydonium* and their deduced aa sequences used for phylogenetic analyses of this subfamily of kinases (Kruse et al. 1996). The first sequence, GCSTKn, displays a pseudosubstrate segment, the typical phorbol esters/diacylglycerol-binding domain, two zinc fingers, and the Ser/Thr kinase catalytic domain (Hanks and Quinn 1991). Hence, GCSTKn has to be grouped with the "novel" (Ca^{2+}-independent) protein kinase C (PKC; Stabel and Parker 1991; Hardie and Hanks 1995). The second kinase, GCSTKc, consists of the pseudosubstrate segment, the typical phorbol esters/diacylglycerol-binding domain, two zinc fingers, an ATP/GTP-binding site motif, the C2 domain signature, which is probably involved in Ca^{2+} binding, and the Ser/Thr kinase catalytic domain. This molecule belongs to the "conventional" (Ca^{2+}-dependent) protein kinases C (Stabel and Parker 1991; Hardie and Hanks 1995; Müller 1997).

A phylogenetic comparison of the Ser/Thr kinase catalytic domains of two sponge enzymes with the respective domains from other known sequences has been published (Kruse et al. 1996). The results revealed that the metazoan PKCs obtained from *Homo sapiens*, *Rattus norvegicus*, *Lytechinus pictus*, *Drosophila melanogaster*, *Caenorhabditis elegans*, and *G. cydonium* are among themselves closely related and show only lower homology to the Ser/Thr kinases of the protozoan *Trypanosoma brucei*, yeast *Saccharomyces cerevisiae*, the plant *Zea mays*, and the bacterial organism *Myxococcus xanthus* (Fig. 8).

5.3.2 *Tyr Kinases*

Until now, no protein with a TK domain has been described from Protozoa or Bacteria. The sponge RTK includes a TK domain which shares the characteristic features known from those of other metazoan enzymes (Hardie and Hanks 1995; Müller and Schäcke 1996). High homology in the *G. cydonium* TK domain was found in nerve growth-factor receptors (*TRKB*_MOUSE and *TRKA*_HOMO), RTK genes of the insulin-receptor subfamily (IRR_HOMO), and related enzymes (SEV_DROSO); in addition, homologies were found in the fibroblast growth-factor subfamily of RTKs (FGR1_HOMO) and several nonreceptor TKs, e.g., the TK domains of the nonreceptor TK genes from the Abl subfamily (ABL_HOMO) and other selected nonreceptor TKs (TEC_DROSO and FER_HOMO) (Gamulin et al. 1997; Fig. 9).

The phylogenetic tree based on the alignment of TK domains of tyrosine kinases shows that all RTKs used for the dendrogram fall in one branch of the tree while the nonreceptor TKs are grouped in a second one. The sponge RTK is placed in a separate branch, which splits off first from the common tree of metazoan kinases (Gamulin et al. 1997; Fig. 8).

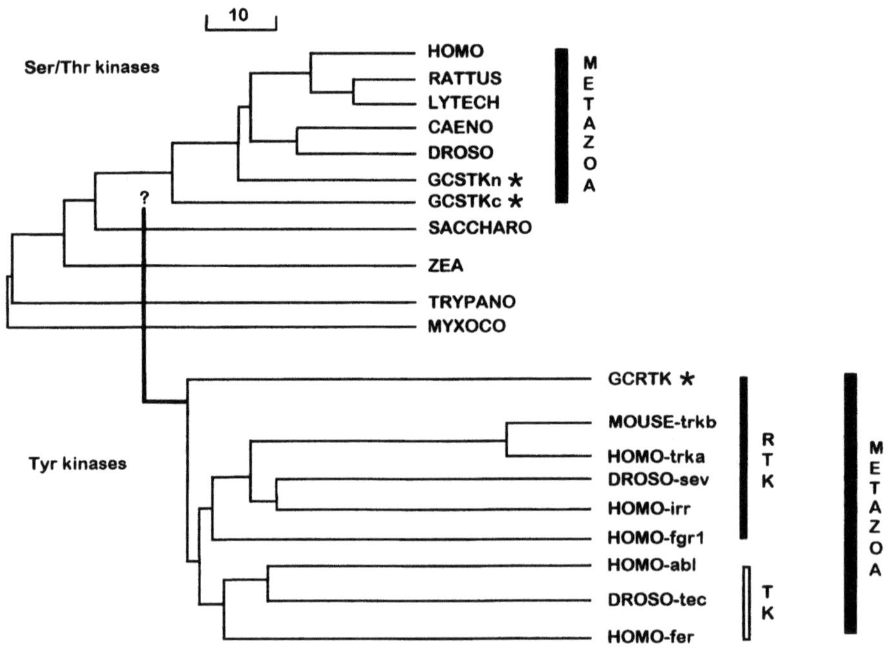

Fig. 8. Phylogenetic tree based on similarity-matrix analysis of the *Ser/Thr kinase* domain (Kruse et al. 1996) present in sequences of the following organisms. Metazoa: *HOMO* (*Homo sapiens*), *RATTUS* (*Rattus norvegicus*), the sea urchin *LYTECH* (*Lytechinus pictus*), *CAENO* (*Caenorhabditis elegans*), *DROSO* (*Drosophila melanogaster*) and the sponge *GCSTKn* and *GCSTKc* (*Geodia cydonium*); Fungi: *SACCHARO* (*Saccharomyces cerevisiae*); Planta: *ZEA* (*Zea mays*); Protozoa: *TRYPANO* (*Trypanosoma brucei*) and Bacteria: *MYXOCO* (*Myxococcus xanthus*). Based on the high sequence homology of the catalytic domains of the sponge, *G. cydonium*, Ser/Thr kinases (*GCSTKn* and *GCSTKc*) and the Tyr kinase (*GCRTK*; see Fig. 9), a relationship between these two groups of kinases is proposed. *Tyr kinase domain* (Gamulin et al. 1997): tree deduced from the multiple alignment of the Tyr kinase domain of RTK gene of *G. cydonium* (*GCRTK*) with those of (1) the following *receptor Tyr kinases* (*RTK*): neurotrophin-4 receptor (*MOUSE-trkb*), nerve growth factor receptor (*HOMO-trka*), *Drosophila melanogaster* sevenless receptor protein tyrosine kinase (*DROSO-ser*), insulin receptor-related receptor (*HOMO-irr*), and basic fibroblast growth factor receptor 1 precursor (*HOMO-fgr1*) as well as (2) the *non-receptor Tyr kinases* (*TK*): ABL protooncogene tyrosine kinase (*HOMO-abl*), *Drosophila melanogaster* SRC protein tyrosine kinase (*DROSO-tec*), and FES/FPS-related protein tyrosine kinase (*HOMO-ter*). The sequences selected are those displaying highest homology to the sponge kinase. The *G. cydonium* sequences are marked with an asterisk. For sequence comparison the CLUSTAL program was applied; the relationships based on the number of similar aa between the species are given in matrix units

5.3.3
Relationship Between Ser/Thr and Tyr Kinases

The close homologies between vertebrate and higher invertebrate Ser/Thr and Tyr kinases are well established (Hunter 1991). With the isolation of these from one member of the lowest metazoan phylum, the sponges, a direct comparison between the catalytic domains of these sequences was possible (Fig. 9).

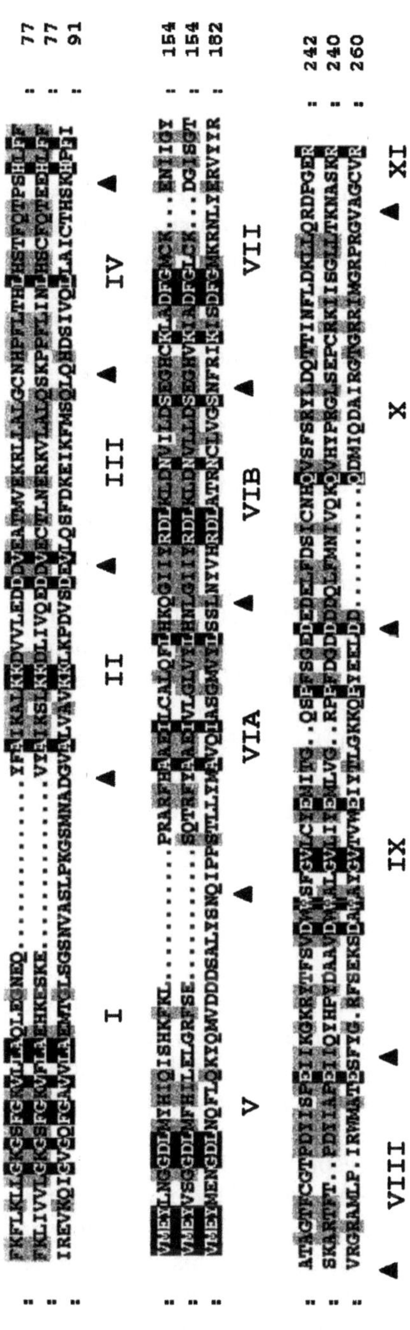

Fig. 9. Alignment of the deduced aa sequence of the catalytic domains of the sponge Ser/Thr kinases (*GCSTKn*, *GCSTKc*) as well as the corresponding domain in RTK (*GCRTK*). The delineation of the *subdomains I-XI* within the catalytic domains is adopted from Hardie and Hanks (1995). Residues conserved among all three sequences are shown in *inverted type*, while residues conserved in two of the sequences are *shaded*

As with the catalytic domains of the kinases from the higher Metazoa, the sponge domains could also be further subdivided into 12 smaller subdomains (Hardie and Hanks 1995). This holds true for both the sponge Ser/Thr kinases (GCSTKn and GCSTKc) and the Tyr kinases (RTK). The alignment, displaying high homology between the three TK domains of the sponge kinases, is shown in Fig. 9. In subdomain I, the ATP-binding site (consensus: GxGxxGxV) and in subdomain II, Lys (K), which is required for kinase activity, are present; this Lys is very likely involved in the phosphotransfer reaction, possibly functioning in proton transfer (Russo et al. 1985). The aa D (Asp) and N (Asn) in subdomain VIB as well as the DFG aa triplet in subdomain VII are present in the sponge sequence; DFG has been implicated in ATP binding (Hanks et al. 1988). This triplet is the most conserved portion in the catalytic domain. Asp (D) in subdomain VIB as well as Asp (D) in subdomain VII are thought to interact with the phosphate groups of ATP via a Mg^{2+} salt bridge (Brenner 1987). Signatures in subdomains III, IV, V, VIA, VIII, IX, X, and XI are in general less well conserved (Hardie and Hanks 1995).

The characteristic features of the catalytic domains of the Ser/Thr kinases are not restricted to Metazoa but are also found in other eukaryotes and also in Bacteria (Kruse et al. 1996). In contrast, catalytic domains which are characteristic for Tyr kinases have been isolated in Metazoa only. However, based on the obvious relationship both with respect to aa identity and division into subdomains among the three catalytic domains of the sponge kinases (Fig. 9), it is conceivable that the evolution of the Ser/Thr kinases to the Tyr kinases happened just before the appearance of the sponges. This relationship is indicated in Fig. 8.

6
Directed Evolution?

As a result, it can be deduced from the examples mentioned above that one major event which allowed evolution of Protozoa to Metazoa was the invention of the genetic apparatus to insert introns and hence to facilitate the creation of mobile modules (Patthy 1995), thus facilitating the *subsequent* process of exon shuffling (Gilbert 1978).

With the isolation of the first sponge genes, especially the one coding for RTK, it has been established that modular proteins, created by exon shuffling, are common to *all* metazoan phyla. This mechanism of exon shuffling is apparently absent in plants and protists (Patthy 1995). If this view can be accepted, then the "burst of evolutionary creativity" (Patthy 1995) during the period of Cambrian explosion which resulted in the big bang of metazoan radiation (Lipps and Signor 1992; Kerr 1993, 1994; Morris 1993) was driven by the process of modularization. During this process, the already existing domains were transformed into mobile modules allowing the composition of mosaic proteins (Patthy 1994; Fig. 4).

For example, the Ig-like domain is not an invention of Metazoa. Molecules featuring Ig-like domains appeared early in eukaryotic evolution; e.g., they are present in yeast α-agglutinin cell wall-associated protein (Chen et al. 1995). However, their use as modules, as building blocks, for the creation of mosaic proteins became possible only after the new step in evolution was acquired which allowed exon shuffling. The mechanism of modularization is more universal and more versatile – it can be applied for all already existing domains – than the process of forming of new domains. Therefore, it can be assumed that during the transition from Protozoa to Metazoa, a process which lasted approximately 1000 million years (Doolittle et al. 1996), the formation of domains with distinct folds was at the center of evolution. After having reached a critical number of domains, the mechanism of modularization allowed a rapid formation of a series of mosaic proteins by exon shuffling.

The formation of domains was subjected to chemical and physical constraints (Goodsell and Olson 1993) as well as to direct functional optimization. Within these frames the creation of domains during the evolution of Protozoa was accidental. After the introduction of mobile modules, which used the already existing domains as building blocks for mosaic proteins, the different metazoan phyla could evolve.

Acknowledgements. I thank Dr. D. Weinblum (Institut für Physiologische Chemie, Universität Mainz) and Dr. L. Patthy (Biological Research Center, Budapest, Hungary) for their valuable discussions. Support by the Deutsche Forschungsgemeinschaft (Mü 348/13-1), the Stiftung Volkswagenwerk, the Fonds der Chemischen Industrie (0161191), and the International Human Frontier Science Program (RG-333/96-M) is gratefully acknowledged.

References

Baldauf SL, Palmer JD (1993) Animals and fungi are each other's closest relatives: congruent evidence from multiple proteins. Proc Natl Acad Sci USA 90: 11558–11562

Bardwell JCA, Craig EA (1984) Major heat shock of *Drosophila* and the *Escherichia coli* heat-inducible DNAk genes are homologous. Proc Natl Acad Sci USA 81: 848–852

Barnes RD (1980) Invertebrate zoology, 4th edn. Saunders, Philadelphia

Baron M, Norman DG, Campbell ID (1991) Protein molecules. Trends Biochem Sci 16: 13–17

Berger SM, Moore C, Sharp PA (1997) Spliced segments at the 5′ terminus of adenovirus 2 late mRNA. Proc Natl Acad Sci USA 74: 3171–3175

Blake CCF (1978) Do genes-in-pieces imply proteins-in-pieces? Nature 273: 267

Brenner S (1987) Phosphotransferase sequence homology. Nature 329: 1

Cavalier-Smith T (1985) Eukaryotic gene number, non-coding DNA and genome size. In: Cavalier-Smith T (ed) The evolution of genome size. Wiley, New York

Cavalier-Smith T (1991) Intron phylogeny: a new hypothesis. Trends Genet 7: 145–148

Chen MH, Shen ZM, Bobin S, Kahn PC, Lipke PN (1995) Structure of *Saccharomyces cerevisia* α-agglutinin: evidence for a yeast cell wall protein with multiple immunoglobulin-like domains with atypical disulfides. J Biol Chem 270: 26168–26177

Chothia C (1992) One thousand families for the molecular biologist. Nature 357: 543–544

Cooper EL, Rinkevich B, Uhlenbruck G, Valembois P (1992) Invertebrate immunity: another viewpoint. Scand J Immunol 35: 247–266

Darnell JE, Doolittle WF (1986) Speculations on the early course of evolution. Proc Natl Acad Sci USA 83: 1271–1275

Doolittle RF, Feng DF, Tsang S, Cho G, Little E (1996) Determining divergence times of the major kingdoms of living organisms with a protein clock. Science 271: 470–477

Gamulin V, Skorokhod A, Kavsan V, Müller IM, Müller WEG (1997) Experimental indication against blockwise evolution of metazoan protein molecules: example, receptor tyrosine kinase gene from the sponge *Geodia cydonium*. J Mol Evol 44: 242–252

Geer P vd, Hunter T, Lindberg RA (1994) Receptor protein-tryosine kinases and their signal transduction pathways. Annu Rev Cell Biol 10: 251–337

Gilbert W (1978) Why genes in pieces? Nature 271: 501

Goodsell DS, Olson AJ (1993) Soluble proteins: size, shape and function. Trends Biochem Sci 18: 65–68

Gupta RS, Singh B (1994) Phylogenetic analysis of 70kD heat shock protein sequences suggests a chimeric origin for the eukaryotic cell nucleus. Curr Biol 4: 1104–1114

Gutierrez G, Oliver JL, Marin A (1993) Dinucleotides and G + C content in human genes: opposite behaviour of GpG, GpC, and TpC at II–III codon positions and in introns. J Mol Evol 27: 131–136

Hanks SK, Quinn AM (1991) Protein kinase catalytic domain sequence database: identification of conserved features of primary structure and classification of family members. Methods Enzymol 200A: 3–37

Hanks SK, Quinn AM, Hunter T (1988) The protein kinase family: conserved features and deduced phylogeny of the catalytic domains. Science 241: 42–52

Hardie G, Hanks S (1995) The protein kinase facts book: protein-tyrosine kinases. Academic Press, London

Hensel R, Zwickl P, Fabry S, Lang J, Palm P (1989) Sequence comparison of glyceraldehyde-3-phosphate dehydrogenase from the three urkingdoms: evolutionary implication. Can J Microbiol 35: 81–85

Hildemann WH (1981) Immunophylogeny: from sponges to hagfish to mice. In: Hildemann WH (ed) Frontiers in immunogenetics. Elsevier, North Holland, New York, pp 3–19

Humphreys T, Reinherz EL (1994) Invertebrate immune recognition, natural immunity and the evolution of positive selection. Immunol Today 15: 316–320

Hunkapiller T, Goverman J, Koop BF, Hood L (1989) Implications of the diversity of the immunoglobulin gene superfamily. Cold Spring Harbor Symp Quant Biol 54: 15–29

Hunter T (1991) Protein kinase classification. Methods Enzymol 200: 3–81

Jones MD, Petersen TE, Nielsen KM, Magnusson S, Sottrup-Jensen L, Gausing K, Clark BFC (1980) The complete amino-acid sequence of elongation factor Tu from *Escherichia coli*. Eur J Biochem 108: 507–526

Kerr RA (1993) Evolution's big bang gets even more explosive. Science 261: 1274–1275

Kerr RA (1994) Timing evolution's early bursts. Science 267: 33–34

Knoll AH (1994) Proterozoic and Early Cambrian protists: evidence for accelerating evolutionary tempo. Proc Natl Acad Sci USA 91: 6743–6750

Konishi J, Denda K, Oshima T, Wakagi T, Uchida E, Ohsumi Y, Anraku Y, Matsumoto T, Wakabayashi T, Mukohata Y, Ihara K, Inatomi K, Kato K, Ohta T, Allison WS, Yoshida M (1990) Archaebacterial ATPases: relationship to other ion-translocating ATPase families examined in terms of immunological cross-reactivity. J Biochem 108: 554–559

Koziol C, Wagner-Hülsmann C, Mikoc A, Gamulin V, Kruse M, Pancer Z, Schäcke H, Müller WEG (1996) Cloning of a heat-inducible biomarker, the cDNA encoding the 70-kDa heat shock protein, from the marine sponge *Geodia cydonium*: response to natural stressors. Mar Ecol Prog Ser 136: 153–161

Koziol C, Kruse M, Batel R, Badria FA, Müller IM, Müller WEG (1997) Characterization of cDNAs encoding DnaJ from the marine sponge *Geodia cydonium* increased gene expression in response to heat shock. Mar Ecol Progr Ser: in press

Kruse M, Gamulin V, Cetkovic H, Pancer Z, Müller IM, Müller WEG (1996) Molecular evolution of the metazoan protein kinase C multigene family. J Mol Evol 43: 374–383

Lazcano A (1994) The RNA world, its predecessors, and its descendants. In: Bengtson S (ed) Early life on earth. Columbia University Press, New York, pp 70–80

Lazcano A, Guerrero R, Margulis L, Oro J (1988) The evolutionary transition from RNA to DNA in early cells. J Mol Evol 27: 283–290
Lechner K, Heller K, Boeck A (1988) Gene for the diphtheria toxin-susceptible elongation factor 2 from *Methanoccocus vannielli.* Nucleic Acids Res 16: 7817–7826
Lipps JH, Signor PW (eds) (1992) Origin and early evolution of Metazoa. Plenum Press, New York
Long M, de Souza S, Gilbert W (1995) Evolution of the intron-exon structure of eukaryotic genes. Curr Opin Genet Dev 5: 774–778
Macario AJL, Dugan CB, Conway de Macario E (1991) A DnaK homolog in the archaebacterium *Methanosarcina mazei* S6. Gene 108: 133–137
Macario AJL, Dugan CB, Clarens M, Conway de Macario E (1993) DnaJ in Archaea. Nucleic Acids Res 21: 2773
Marchalonis JJ, Schluter SF (1990) On the relevance of invertebrate recognition and defense mechanisms to the emergence of the immune response of vertebrates. Scand J Immunol 32: 13–20
Mattick JS (1994) Introns: evolution and function. Curr Opin Genet Dev 4: 823–831
Mehl D, Müller I, Müller WEG (1997) Molecular biological and palaeontological evidence that Eumetazoa, including Porifera (sponges), are of monophyletic origin. In: Proc Int Conf on Sponge science. 12–16 March 1996, Otsu, Japan, Springer Berlin Heidelberg New York (in press)
Morris SC (1993) The fossil record and the early evolution of the Metazoa. Nature 361: 219–225
Müller WEG (1995) Molecular phylogeny of Metazoa [animals]: monophyletic origin. Naturwiss enschaft 82: 321–329
Müller WEG (1997) Molecular phylogeny of Eumetazoa: experimental evidence for monophyly of animals based on genes in sponges [Porifera]. Prog Mol Subcell Biol (in press)
Müller WEG, Schäcke H (1996) Characterization of the receptor protein-tyrosine kinase gene from the marine sponge *Geodia cydonium.* Prog Mol Subcell Biol 17: 183–208
Müller WEG, Müller IM, Gamulin V (1994) On the monophyletic evolution of the Metazoa. Braz J Med Biol Res 27: 2083–2096
Ohki M, Tamura F, Nishimura S, Uchida H (1986) Nucleotide sequence of the *Escherichia coli* DnaJ gene and purification of the gene product. J Biol Chem 261: 1778–1781
Orengo CA, Jones DT, Thornton J (1994) Protein superfamilies and domain superfolds. Nature 372: 631–634
Palmer JD, Logsdon JM (1991) The recent origins of introns. Curr Opin Genet Dev 1: 470–477
Pancer Z, Cooper EL, Müller WEG (1996a) A tunicate (*Botryllus schlosseri*) cDNA reveals similarity to vertebrate antigen receptors. Immunogenetics 45: 69–72
Pancer Z, Scheffer U, Müller I, Müller WEG (1996b) Cloning of sponge (*Geodia cydonium*) and tunicate (*Botryllus schlosseri*) proteasome subunit epsilon (PRCE): implications about the evolution of the vertebrate MHC-encoded homologue LMP7 (PRCC). Biochem Biophys Res Commun 228: 406–410
Pancer Z, Diehl-Seifert B, Rinkevich B, Kruse M, Blumbach B, Müller WEG. (1997) Cloning of a novel putative tunicate (*Botryllus schlosseri*) mannose-binding protein that features an Ig-like domain DNA Cell Biol (in press)
Patthy L (1987) Intron-dependent evolution: preferred types of exons and introns. FEBS Lett 214: 1–7
Patthy L (1994) Introns and exons. Curr Opin Struct Biol 4: 383–392
Patthy L (1995) Protein evolution by exon-shuffling. Springer Berlin Heidelberg, New York
Reiter WD, Hüdepohl U, Zillig W (1990) Mutational analysis of an archebacterial promoter: essential role of a TATA box for transcription efficiency and start-site selection in vivo. Proc Natl Acad Sci USA 87: 9509–9513
Reitner J, Mehl D (1995) Early paleozoic diversification of sponges: new data and evidences. Geol Paläont Mitt Innsbruck 20: 335–347
Rozas J, Rozas R (1995) DnaSP, DNA sequence polymorphism: an interactive program for estimating population genetics parameters from DNA sequence data. Comput Appl Biosci 11: 621–625

Russo WM, Lukas TJ, Cohen S, Staros VJ (1985) Identification of residues in the nucleotide binding site of the epidermal growth factor receptor-kinase. J Biol Chem 260: 5205–5208
Schäcke H, Schröder HC, Gamulin V, Rinkevich B, Müller IM, Müller WEG (1994a) Molecular cloning of a tyrosine kinase gene from the marine sponge *Geodia cydonium*: a new member belonging to the receptor tyrosine kinase class II family. Mol Membr Biol 11: 101–107
Schäcke H, Schröder HC, Müller IM, Müller WEG, Gamulin V, Rinkevich B (1994b) The immunoglobulin superfamily includes members from the lowest invertebrates to the highest vertebrates. Immunol Today 15: 497–498
Seino S, Seino M, Nishi S, Bell GI (1989) Structure of the human insulin receptor gene and characterization of its promoter. Proc Natl Acad Sci USA 86: 114–118
Sima P, Vetvicka V (1993) Evolution of immune reaction. Crit Rev Immunol 13: 83–114
Smith LC, Davidson EH (1992) The echinoid immune system and the phylogenetic occurrence of immune mechanisms in deuterostomes. Immunol Today 13: 356–361
Stabel S, Parker PJ (1991) Protein kinase C. Pharmacol Ther 51: 71–95
Steiner M, Mehl D, Reitner J, Erdtmann BD (1993) Oldest entirely preserved sponges and other fossils from the Lowermost Cambrian and a new facies reconstruction of the Yangtze Platform (China). Berl Geowiss Abh (E) 9: 293–329
Sun SC, Lindstroem I, Boman HG, Faye I, Schmidt O (1990) Hemolin: an insect-immune protein belonging to the immunoglobulin superfamily. Science 250: 1729–1732
Suzuki T, Takagi T, Furukohri T, Kawamura K, Nakauchi M (1990) A calcium-dependent galactose-binding lectin from the tunicate *Polyandocarpa misakiensis*. Isolation, characterization, and a mini acid sequence. J Biol Chem 265: 1274–1281
Van de Vyver G (1988) Histocompatibility responses in freshwater sponges: a model for studies of cell-cell interactions in natural populations and experimental systems. In: Grosberg RK, Hedgecock D, Nelson K (eds) Invertebrate historecognition. Plenum Press, New York, pp 1–14
Walter M (1996) Old fossils could be fractal frauds. Nature 383: 385–386
Walter MR (1994) Stromatolites: the main geological source of information on the evolution of the early benthos. In: Bengtson S (ed) Early life on earth. Columbia University Press, New York, pp 270–286
Williams AF, Barclay AN (1988) The immunoglobulin superfamily - domains for cell surface recognition. Annu Rev Immunol 6: 381–405
Williamson B (1977) DNA insertions and gene structure. Nature 270: 295–297
Woese CR (1987) Bacterial evolution. Microbiol Rev 51: 221–271
Woese CR, Achenbach L, Rouviere P, Mandelco L (1991) Archaeal phylogeny: reexamination of the phylogenetic position of *Archaeoglobus fulgidus* in light of certain composition-induced artifacts. Syst Appl Microbiol 14: 364–371
Yamane M, Hattori H, Sugito K, Hayshi Y, Tohnai I, Ueda M, Nishizawa K, Ohtsuka K (1995) Cotranslation and colocalization of hsp40 (DnaJ) with hsp70 (DnaK) in mammalian cells. Cell Struct Funct 20: 157–166
Zhao L, Kansot MR (1996) In search of a function for hemolin, a hemolymph protein from the immunoglobulin superfamily. J Insect Physiol 42: 73–79
Ziegler SF, Bird TA, Schneringer JA, Schooley KA, Baum PR (1993) Molecular cloning and characterization of a novel receptor protein tyrosine kinase from human placenta. Oncogene 8: 663–670
Zillig W (1991) Comparative biochemistry of Archaea and Bacteria. Curr Biol 1: 544–551

Individuality in Early Eukaryotes and the Consequences for Matazoan Development

Neil W. Blackstone

1 Introduction

The question of the monophyly of the metazoans clearly requires elucidating characters which distinguish the metazoans from their eukaryotic sister groups. Such characters will likely be diverse and include sequence data (e.g., Wainright et al. 1993; Doolittle et al. 1996) and additional biochemical and cellular characters (e.g., collagen, chitin, see Wilmer 1990; basal membranes, gap junctions, nematocysts, colloblasts, etc., see Schram 1991; Backeljau et al. 1993; Schram and Ellis 1994; extracellular matrix characteristics, see Morris 1993). Thus, to a large extent, the question of the monophyly of the metazoans will be resolved by characters which emerge from the cellular and subcellular natural histories of metazoans and other eukaryotes. A more complete understanding of this question will remain elusive as long as the selective forces which shaped these natural histories remain unexplored (see Buss 1987; Blackstone 1995; Coyne 1996). I therefore propose that discussions of the monophyly of the metazoans should not just elucidate the biochemical and molecular shared derived characters for this group, but should consider the Darwinian evolution of these characters as well.

It would be an understatement to say that this is not a simple task. Key evolutionary events took place many hundreds of millions of years ago (e.g., Knoll 1992), and transitional forms can be expected to have become extinct long ago (Maynard Smith and Szathmary 1995; Bonner 1996). Furthermore, despite stunning progress in the last 50 years (e.g., Chance 1991; Lawrence 1992; Bonner 1996), the cellular and subcellular natural histories of metazonans and other eukaryotes are just beginning to be elucidated. Thus, attempts to trace the selective forces molding particular character states is perhaps premature; rather, it is first necessary to develop a general framework which will subsequently allow more focused considerations. A useful first step in this task is to clarify the actions of natural selection on cellular and molecular characters. In its essential form, Darwinian evolution will occur when a biological unit displays heritable variation which is subject to natural selection (Lewontin 1970). Thus genes, organelles, cells, and of course multicellular

[1] Department of Biological Sciences, Northern Illinois University, Dekalb, Illinois 60115, USA

Progress in Molecular and Subcellular Biology, Vol. 19
W.E.G. Müller (Ed.)

individuals can all, under certain circumstances, simultaneously act as units of evolution. Furthermore, the history of life is a history of an elaboration of these units of evolution: genes within cells, cells within cells, and cells within multicellular individuals (Buss 1987). At each step in this elaboration of the biological units of evolution, novel selective scenarios are expected to prevail. For instance, consider a prokaryotic cell becoming an endosymbiotic organelle within a eukaryotic host. Prior to becoming an endosymbiont, such a cell would be selected by the external environment alone. However, subsequent to becoming an obligate endosymbiont, the prokaryotic cell would now be selected by traits expressed by the host cell (i.e., the success of the host cell would now determine the success of the symbiont). Once engaged in the symbiosis, variant endosymbionts could influence not only their own replication rate, but the replication rate of their hosts. Further, the host could develop mechanisms to constrain the expression of variation of the endosymbiont. A units-of-evolution perspective focuses on such synergistic and antagonistic effects of the two evolutionary units on each other.

Two premises are immediately apparent. First, such an interaction between units of evolution is likely to exert a powerful effect on each unit, resulting in substantial evolutionary change as the selfish interests of each unit are resolved and reconciled to an extent which is sufficient to allow the symbiosis to persist. Second, this resolution and reconciliation must always favor the higher-level unit. If this is not the case, that is, if the fitness of the higher-level unit is compromised in favor of the lower unit, then the higher-level unit will be at a competitive disadvantage relative to other comparable units in the environment, and natural selection will diminish the representation of the compromised higher-level unit (and its associated population of lower-level units) in the population (Fig. 1 exemplifies this point). The history of life no doubt contains a multitude of such evolutionary events (e.g., Siddall et al. 1995).

This brief discussion of natural selection suggests that a significant portion of evolutionary change will be compressed into the relatively brief time periods in which transitions between units of evolution occur. Considerable insight into the cellular and molecular natural histories of modern metazoans may thus be gained by considering some of the key transitions between units of evolution. In the context of considering the monophyly of the metazoans, two key transitions are expected to be of particular importance: the transition from simple cells to complex cells (i.e., cells within cells; Blackstone 1995; Szathmary and Maynard Smith 1995) and the transition from single cells to multicellular organisms (Buss 1987; Szathmary and Maynard Smith 1995; Bonnner 1996).

It is during the latter transition that many of the homologous features of metazoan development likely evolved, assuming that the evolution of multicellularity in metazoans was a unique event (i.e., that it was a separate event from the evolution of multicellularity in other groups). For instance, it was at this time that the shared derived characteristics of the metazoan

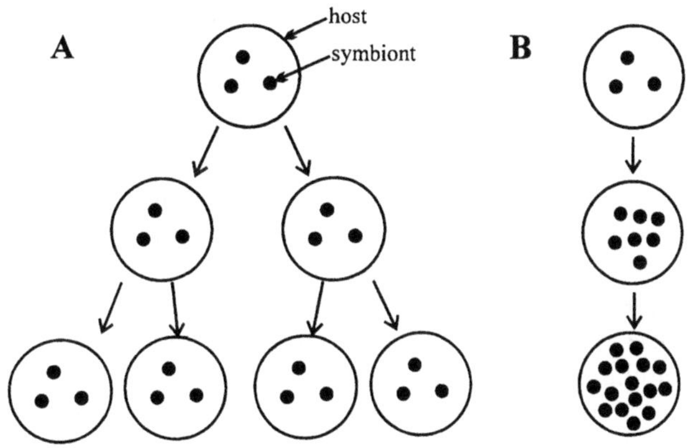

Fig. 1A,B. Diagrammatic representation of the primacy of the fitness of the higher-level unit in determining the success of a lineage made up of two units of evolution, in this case a host cell and an obligate symbiont. In lineage **A**, only sufficient energy is allocated into the replication of the symbiont to allow its persistence, while the remaining energy is allocated into the replication of the host. In lineage **B**, more energy is allocated into the replication of the symbiont, and little energy remains to be allocated into the replication of the host. The symbiont in lineage **B** has a high fitness in the internal environment within the host, but its fitness in the external environment is low because its host loses in competition to the host of lineage **A**

extracellular matrix (Morris 1993) likely evolved. These and other complex biochemical mechanisms are more likely to be phylogenetically informative than single molecules, since structural or functional constraints upon the latter may mask convergence (Morris and CoBabe 1991; Morris 1993). Consequently, I predict that elucidating the specific mechanisms of metazoan development will also provide a rich array of phylogenetically informative characters for this group.

This discussion will exemplify this prediction by focusing on a subset of these mechanisms of metazoan development, in particular, those mechanisms that express the relationship between metabolism and development. While theories of the metabolic control of development have long been in disrepute (Mitman and Fausto-Sterling 1992), the weight of recent empirical evidence has caused a reawakening of these ideas (e.g., Newman 1973; Allen and Balin 1989; Adair et al. 1990; Nijhout 1990; Fanburg et al. 1992), and at present the amount of information on this subject is expanding rapidly. I will first review the general subject of the relationship between metabolism and development in metazoans, and particular examples will be presented. Subsequently, I will suggest that evolutionary considerations of two major transitions in the history of life (simple to complex cells and single cells to multicellular organisms) in fact predict the existence of a relationship between metabolism and development. I propose that the circumstances surrounding the evolution of modern eukaryotic cells may have dictated certain evolutionary events; these events, once transpired, may subsequently have channeled the course of eukaryotic evolution, including that of the metazoans at the time of the transition from unicellularity to multicellularity. Thus, while the putative shared derived characters relating metazoan development to metabolism likely evolved during the transition from unicellularity to multicellularity, I suggest that a predisposition for this relationship evolved earlier, at the time of the transition from simple to complex cells. The same logic suggests that a relationship between metabolism and development may be found in other multicellular eukaryotes (particularly plants and fungi), but I predict that the mechanistic implementation of this relationship will differ between these groups and that these specific mechanisms will be shared derived characters for each multicellular group.

The bulk of this discussion will thus be somewhat removed from issues of systematics in general and metazoan monophyly in particular. Nevertheless, this elaboration of my views on the evolution of metabolism and development is done with the intent of emphasizing the importance of Darwinian evolution to issues of phylogenetic systematics, both in general and in particular. Subsequent to developing this elaborate example, I will return to considerations of the monophyly of the metazoans in the conclusions.

2 The Evolution of Metabolism and Development

Metazoan development is a derived characteristic, and it is therefore expected that features of metazoan development will be informative in terms of this evolutionary history (Buss 1987). This is particularly true for the relationship between metabolism and development. I will first review some old ideas and new data which provide support for this relationship, and then I will outline the evolutionary logic of this relationship.

2.1 Metabolism and Development, Revisited

In many ways, the publication of C.M. Child's *Patterns and Problems in Development* (1941) marked the zenith of theories linking "metabolic activity" to organismal development. Subsequently, these theories were entirely supplanted by the theory of variable gene activity (e.g., Morgan 1934; Davidson 1968). While metabolic activity clearly subsumes gene activity, Child and like-minded colleagues commonly focused on energy metabolism and its manifestation (e.g., oxygen uptake). Energy metabolism was thus viewed as guiding gene activity, rather than the reverse. The contrast to some modern viewpoints is thus sharp (see Nijhout 1990). Nevertheless, various aspects of recent research increasingly suggest that some of Child's views may provide an understanding of development that is presently lacking (e.g., Newman 1973; Allen and Balin 1989; Adair et al. 1990; Nijhout 1990; Fanburg et al. 1992; Blackstone 1995; Jouaville et al. 1995; Thomas et al. 1996).

The demise of Child's theories was in part sociological (i.e., the result of social conflicts between groups of scientists who championed different model systems; see Mitman and Fausto-Sterling 1992) and in part due to legitimate scientific concerns. Child and his collaborators first demonstrated metabolic gradients with a number of different kinds of metabolic toxins, then with redox sensitive dyes, and finally by cutting tissue up and measuring the oxygen uptake of each isolated part using Winkler titrations (Child 1941; for additional review and comments, see Allen and Balin 1989, and Bonner 1996). Criticisms of Child have focused on the use of metabolic toxins, on the method of Winkler titrations, and on the inference of oxygen consumption of the intact organism from oxygen consumption of isolated fragments (see discussions of these criticisms in Child 1941). Nevertheless, collectively this work provides unambiguous evidence of metabolic gradients and their association with developmental events. The greatest difficulty with Child's theory was thus one of cause and effect: did metabolic gradients cause developmental gradients, or were the former a consequence of the latter? Clearly, much of development (e.g., differential cell division and movement) will automatically entail energetic differences. Rapidly dividing or moving cells will of necessity consume more energy and oxygen than cells which are relatively quiescent. In modern terms, the essential question is whether metabolic gradients cause variation in gene activity or vice versa. At the time that Child and his collaborators were active, the major impediment to testing this question was the lack of appropriate instrumentation. As discussed by Bonner (1996), the ingenious experiments designed by Child and like-minded colleagues often had competing explanations. For instance, stalks of the hydroid *Tubularia* could be severed from a colony, whereupon one end of the stalk would regenerate a new hydranth. A series of experiments demonstrated that if one end of the stalk were buried in sand, the other end would always regenerate the hydranth, regardless of the original polarity of the stalk. Child suggested that this was caused by gradients of oxygen diffusion, with the buried end receiving less oxygen and consequently the exposed end regenerating the hydranth. However, an equally plausible explanation is that hydranth formation is repressed by a diffusible morphogen emanating from the stalk itself; because of limited diffusion, the buried end maintains a higher concentration of this morphogen and hence cannot form a new hydranth.

It is noteworthy that instrumentation capable of testing many of Child's hypotheses was in the process of being developed toward the end of Child's career (see Chance 1991; Holmes 1992). Of particular interest in this regard are in vivo measures of redox state using mitochondrial spectrofluorescence (e.g., Chance and Baltscheffsky 1958). Even at present, this optical method remains the most sensitive way to assess redox state, and it is capable of distinguishing metabolic levels that confound both positron emission transmission and magnetic resonance spectroscopy (Chance 1991). The history of the idea of metabolic regulation of development might have been quite different if this technique had come to the attention of Child's collaborators earlier.

After a lengthy period of dormancy, theories of the metabolic control of development have now been largely resurrected. Presently, it is widely accepted that oxidative state can have a large impact on gene activity and thus development. For instance, Allen and Balin (1989) provide an extensive review of the "free radical theory of development," and in so doing point out parallels to theories of abnormal development (Ames and Gold 1990a,b) and senescence (Harman 1981). In the same vein, a US National Heart, Lung, and Blood Institute workshop has recently stressed the importance of oxygen and oxidative gradients in modulating gene activity (Fanburg et al. 1992). Numerous other prominent reviews and experimental results can be found (e.g., Schilling and Elliot 1992; Orr and Sohal 1994). To review this vast literature would be beyond the scope of this discussion. Rather, I would like to briefly summarize and comment on several specific reports to elucidate what I consider to be key aspects of this area of research.

2.1.1 *Mitochondria Make Waves*

In eukaryotic cells, the concentration of cytosolic Ca^{2+} controls many cellular activities (e.g., excitability, contraction, exocytosis, growth, metabolism, and gene expression; see Carafoli 1987; Berridge and Irvine 1989; Tsien and Tsien 1990; Allbritton et al. 1992; Bandtlow et al. 1993; Galione 1993). In particular, "second messenger" systems are common in eukaryotes, allowing cell-cell communications in a variety of circumstances (e.g., Berridge and Irvine 1989; Berridge 1993). Typically, stimulation of cell-surface receptors and hydrolysis of a membrane-bound inositol lipid produces at least two second messengers, diacylglycerol and inositol 1,4,5-trisphosphate. The former activates protein kinase C, which phosphorylates a variety of enzymes, while the latter triggers the release of nonmitochondrial calcium stores, thus potentially initiating the variety of events listed above.

Cytosolic and mitochondrial Ca^{2+} are also important regulators of mitochondrial respiration rate (Hansford 1985; Brand and Murphy 1987). There has thus been considerable interest in the possibility of mitochondrial regulation of intracellular calcium and calcium signaling (Nicholls and Akerman 1982). In extant eukaryotes, both the affinity of mitochondria for Ca^{2+} and the rate at which mitochondria import this cation are too low for mitochondria to effectively regulate intracellular calcium concentrations under typical cytosolic conditions (Chance 1965; Hansford 1985; Carafoli 1987). Nevertheless, isolated mitochondria can take up large quantities of Ca^{2+} (e.g., Lehninger 1964), and this may occur in vivo when calcium ions reach abnormally high concentrations (Alberts et al. 1983; Carafoli 1987).

Several recent reports suggest additional complexities. For instance, Jouaville et al. (1995) report that in *Xenopus* oocytes inositol 1,4,5-trisphosphate-induced Ca^{2+} wave activity is strengthened by oxidizable substrates that energize mitochondria. In cells in which mitochondria were metabolically activated, Ca^{2+} wave amplitude, velocity, and interwave period were all increased. These data suggest that potential-driven mitochondrial Ca^{2+} uptake can be a major factor in the regulation of inositol 1,4,5-trisphosphate-induced Ca^{2+} release, thus implicating mitochondrial metabolic state in this ubiquitous second messenger pathway of intracellular Ca^{2+} signaling.

Nevertheless, additional experiments by these authors show that mitochondria fair poorly in competition with sarco-endoplasmic reticulum Ca^{2+}-ATPases; again, both the affinity of mitochondria for Ca^{2+} and the rate at which mitochondria import this cation are too low to allow effective mitochondrial competition. Thus the question remains as to how mitochondria accomplish the observed effects on Ca^{2+} signaling. Possibly, an association between mitochondria and the endoplasmic reticulum occurs, and calcium waves may emanate from this association in these oocytes (Speksnijder et al. 1993). During calcium fluxes, mitochondria in close proximity to the appropriate endoplasmic reticulum should experience a relatively high local concentration of Ca^{2+}, thus overcoming their low Ca^{2+} affinity.

The remaining questions about the mechanisms of mitochondrial effects on intracellular Ca^{2+} should not obscure the implications of the observations and experiments of Jouaville et al. (1995): the metabolic state of mitochondria can potentially mediate the effects of intracellular calcium fluxes and thus modulate many of the cell's activities. No clearer demonstration of the potential for metabolic regulation of development could be expected.

2.1.2
Oxygen Radicals and Cell Signaling

The resurgence of general interest in metabolism and development has been largely due to the very great recent interest in the role of oxygen-free radicals in development (e.g., Chance et al. 1979; Allen and Balin 1989; Fanburg et al. 1992; Schilling and Elliot 1992; Shweiki et al. 1992), carcinogenicity (e.g., Ames and Gold 1990a,b), and senescence (e.g., Chance et al. 1979; Harman 1981; Richter 1988; Floyd 1991; Halliwell and Aruoma 1991; Stadtman 1992; Wallace 1992; Orr and Sohal 1994). In particular, a variety of physiological and developmental phenomena in vascular systems suggest that metabolic and oxidative gradients may produce gradients of gene activity via the effects of oxygen radicals (see Gibbons and Dzau 1996). For instance, links have been suggested between β-amyloid signal peptides and oxygen-free radicals in the vasoactivity of the vertebrate vascular system (Thomas et al. 1996). This peptide may interact with the endothelial cells of blood vessels to produce a local excess of superoxide radicals (O_2^-). Contractions of the blood vessel ensues, possibly because superoxide radicals scavenge the endothelium-derived relaxing factor. Superoxide dismutase, which converts the superoxide anion to hydrogen peroxide (Fig. 2), has the expected protective effects. An oxygen radical-mediated signaling pathway is thus suggested (see also Heinle et al. 1994; Gibbons and Dzau 1996). Such a pathway has specific implications

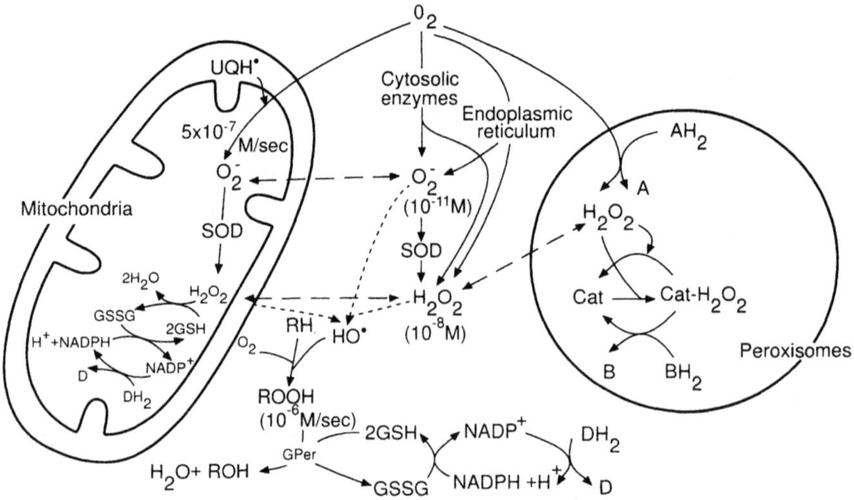

Fig. 2. A general scheme of the roles of catalase (*Cat*), glutathione peroxidase (*GPer*), and superoxide dismutase (*SOD*) in the cytosol, mitochondria, and peroxisomes of the eudaryotic cell. Concentrations and formation rates of oxygen metabolites are estimated. *UGH* · Ubiquinone radical; *GSSG* oxidized glutathione; GSH reduced glutathione; DH_2 and *D* a nonspecified NADP reducing system; *NADPH* and *NADP* nicotinamide adenine dinucleotide phosphate; O_2^- superoxide anion; *HO* · hydroxyl radical; *ROOH* an alkyl hydroperoxide; *B* and BH_2 hydrogen donors of a specificity appropriate to catalase. Cytoplasmic catalase, which is derived from the peroxisome, is not shown. (Chance et al. 1979, with permission)

for the study of human disease, perhaps explaining the significance of deposits of β-amyloid found to be associated with aging and Alzheimer's disease. More general implications include the suggestion of a mechanism by which superoxide radicals can regulate blood flow and thus indirectly affect many aspects of vertebrate physiology, growth, and development. The question remains as to why vertebrates would evolve a signaling pathway with such obvious potential for tissue damage of the sort described by Thomas et al. (1996).

2.1.3 *Oxygen Radicals and Cell Signaling, Continued*

In the eukaryotic cell, metabolism of hydrogen peroxide and the superoxide anion constitutes a small but significant portion of the cell's overall oxygen metabolism (Chance et al. 1979; see Fig. 2). Parallel to studies of superoxide-mediated signaling, there has been considerable interest in hydrogen peroxide-mediated signaling. If signaling with reactive oxygen species was somehow favored by evolution, the later might be expected to be more common, because hydrogen peroxide is less reactive than the superoxide anion. Indeed, Sundaresan et al. (1995) suggest that the effects of the platelet-derived growth factor on vertebrate vascular smooth muscles is mediated by H_2O_2. These growth factors trigger a transient rise in intracellular concentrations of H_2O_2 and have a variety of effects on the smooth muscles relevant to developmental physiology, including tyrosine phosphorylation, mitogen-activated protein kinase stimulation, and DNA synthesis. Both the latter effects and the increase in H_2O_2 can be diminished by increasing the intracellular concentration of catalase or of other chemical antioxidants.

2.1.4
Metabolism and Development in the Basal Metazoa

> That quantitative metabolic differentials seem to be the most conspicuous features in early stages of many forms and that they are essential factors in development is a justifiable conclusion from the data. . . . Undoubtedly, there is interaction between nucleus and cytoplasm, and supposedly different genes become activated or in some way come to play a part in determining the differences in different cells; but . . . differences in gene effect must be based on conditions in the cytoplasm.
>
> (C.M. Child 1941)

It may not be coincidental that Child studied primarily planarians and hydroids (e.g., Child and Hyman 1919), while his nemesis, T.H. Morgan, studied *Drosophila*. Metabolic gradients may play a larger role in the development of these basal metazoans than in the more derived forms, although the evidence to definitively test this hypothesis is not yet available. Nevertheless, modern studies of the development of these animals continue to implicate aspects of metabolism. For instance, Newman (1973) has shown that certain universally occurring metabolites (e.g., ATP) have distinct and repeatable effects on pattern formation in *Hydra*. Further, patterns of colony development in colonial hydroids are sensitive to perturbations of energy allocation and metabolic state (e.g., Fig. 3; Blackstone 1992, in press; Blackstone and Buss 1992, 1993).

2.1.5
Summary

The research summarized above provides a snapshot of current attempts to elucidate the mechanisms of metabolic and oxidative signaling in development. Because of the obvious focus of biomedical research, these mechanisms are best understood in the vertebrates; nevertheless, metabolic signaling may be even more common in the basal metazoans. While the details remain tentative, in broad outline two points are suggested: (1) by-products of oxygen metabolism play a key role in these signaling pathways, and (2) the metabolic state of mitochondria can directly affect such signaling via effects on intracellular calcium fluxes. In eukaryotic cells, oxygen metabolism is largely confined to mitochondria and peroxisomes (Fig. 2), using a broad definition of the latter (see Masters and Crane 1995). Thus these organelles, by regulating

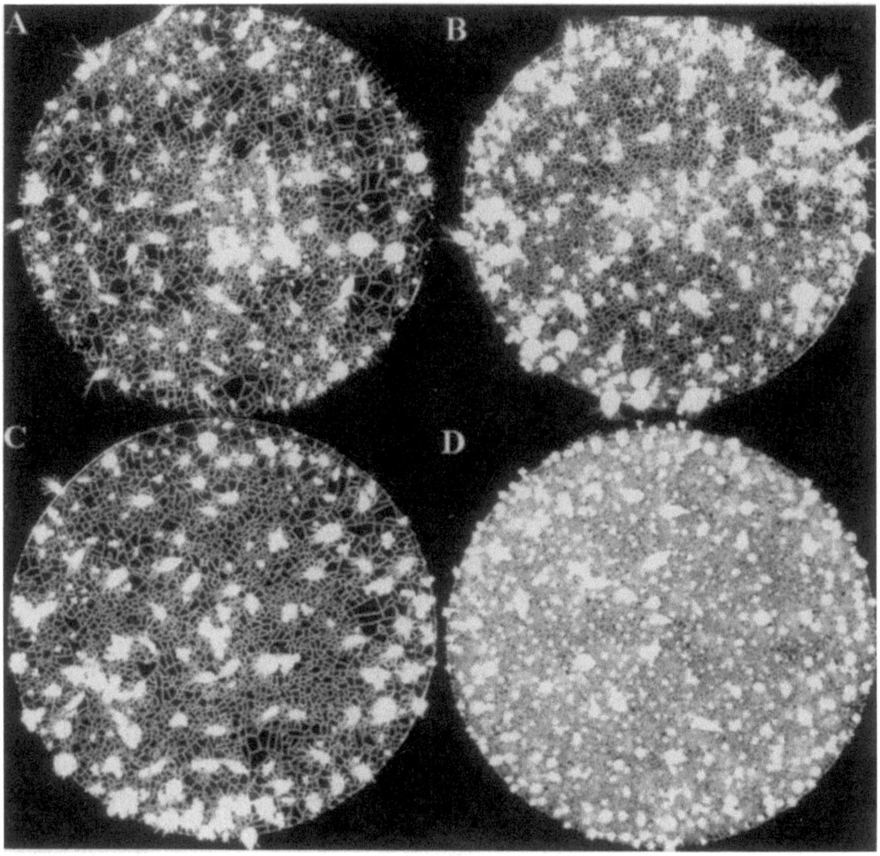

Fig. 3A–D. Background-subtracted images of several genetically identical colonies of the hydroid *Podocoryne carnea* growing on 15-mm diameter glass cover slips. **A** 4-h exposure control; **B** treated with 30 μM dinitrophenol in seawater for 4 h/day; **C** 20-h exposure control; **D** treated with 30 μM dinitrophenol in seawater for 20 h/day. Control colonies have similar morphologies, whereas treated colonies differ dramatically from their controls and from each other. Treatment with dinitrophenol diminishes ATP production, alters the redox state in the direction of oxidation and causes metabolic activation, e.g., an increase in oxygen uptake. (Blackstone, in press)

intracellular concentrations of oxidants and possibly by direct signaling via calcium, can play a key role in the mechanisms of the metabolic regulation of development.

2.2 The Logic of the Metabolic Control of Development

As noted above, the use of reactive oxygen species in the signaling pathways of vertebrates is not without risk. Natural selection at the level of the individual organism would be expected to favor other, less risky signaling pathways. In this context, a units-of-evolution perspective can provide insight into the evolutionary events which may have contributed to the use of oxidants in key signaling pathways. Circumstances surrounding the evolution of modern eukaryotic cells may have dictated certain evolutionary events; these events, once transpired, may subsequently have channeled the evolutionary course of eukaryotes, predisposing them to use reactive oxygen species and other organelle aspects in *intracellular* signaling pathways. At the time of the transition from unicellularity to multicellularity in metazoans, the evolution of a new class of *intercellular* signaling pathways was necessary, and these new pathways could in part be derived from the organelle-based intracellular pathways already in existence. Thus, while the relationship between metazoan development and metabolism likely evolved during the transition from unicellularity to multicellularity, I suggest that a predisposition for this relationship evolved earlier, at the time of the transition from simple to complex cells. These two transitions and their hypothesized effects on signaling pathways are elaborated below.

2.2.1 *Simple to Complex Cells*

One of the major evolutionary transitions in the history of life is from prokaryotes to eukaryotes (Szathmary and Maynard Smith 1995). A major part of this transition is the evolution of cells within cells, i.e., the adoption of an endosymbiotic role by some prokaryotes. For instance, the endosymbiont theory of the origin of the mitochondrion is now generally accepted, largely because of extensive compilations of supporting evidence (e.g., see Baltscheffsky and Baltscheffsky 1981; Margulis 1981; Gray and Doolittle 1982; McCarroll et al. 1983; Spenser et al. 1984; Cavalier-Smith 1987; Gray 1989, 1992; Knoll 1992; Gillham 1994); the theory of the endosymbiotic origin of the peroxisome, while more problematic, also has some support (e.g., de Duve 1969; Cavalier-Smith 1987; Borst 1989; Igual et al. 1992).

From a Darwinian perspective, an endosymbiosis at its inception must have been tenuous at best, with the actions of natural selection at the level of the symbiont constantly threatening to disrupt the integration necessary for the eukaryotic cell to survive (Fig. 1; Williams 1992; Blackstone 1995; Szathmary and Maynard Smith 1995). Possible pathways for units-of-evolution conflicts between the host cell and the protomitochondrion are outlined in Blackstone (1995). The general themes for these conflicts are expected to relate to both energy and oxygen: the aerobic protomitochondrion is expected to have been rich in energy, quite familiar with oxygen, and relatively well supplied with superoxide dismutase and catalase, whereas the anaerobic host cell is thought to have been relatively energy poor and more naive in terms of exposure to oxygen and oxidants.

Subsequent to the establishment of an initial association, natural selection would favor those protomitochondria which could manipulate their host's life history (e.g., the rate of cell division and sexual recombination) to enhance their own replication rate. Given the expected naiveté of the host, obvious candidates for carrying out this manipulation would be quantities of ATP and of oxidants, in particular, either the superoxide anion or hydrogen peroxide, or both (Fig. 4). A variant mitochondrion which could increase the availability of ATP within the host cell could thus increase the host's rate of cell division, thereby providing additional environments for the symbionts (Fig. 4a). In this way, perhaps a risky, host-to-host dispersal stage of the protomitochondrion life history was eliminated and a competitive advantage over other, free-living protomitochondrial lineages was maintained. On the other hand, deleterious mutations resulting from oxidative damage to the host's DNA would have selected for an increase in the frequency with which the host cell engaged in

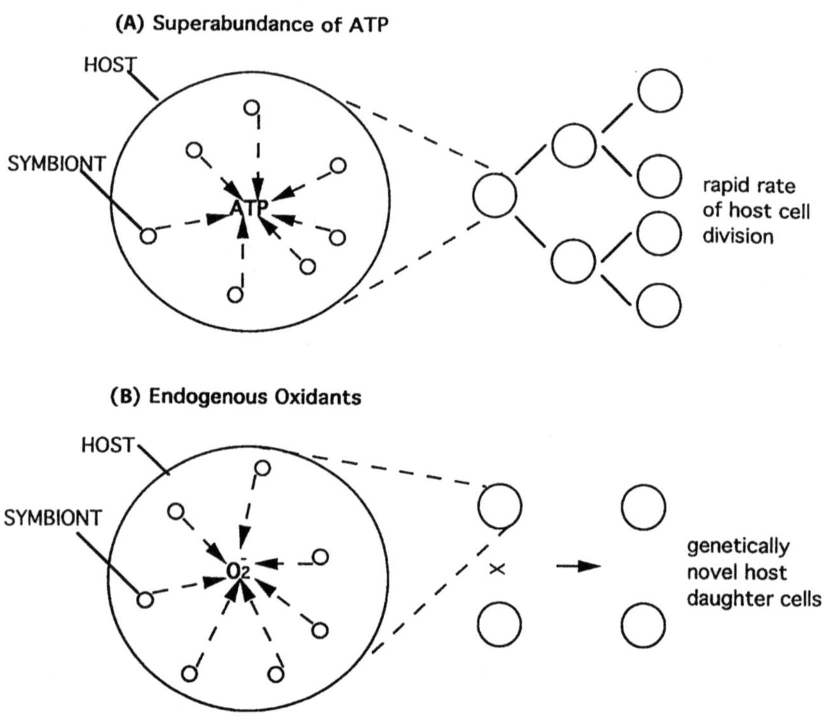

Fig. 4A,B. Possible effects of the endosymbiont on the host include: A the abundance of mitochondrially produced ATP would select for host cells which accelerated their division rates. B The endogenous oxidants formed from mitochondrial O_2^- would cause DNA adducts and mutations in the host cell. The host cell would suffer a higher rate of deleterious mutations, and this would select for higher rates of sexual recombination in the host. Because metabolic activation shifts the redox state in the direction of oxidation, a rapidly dividing host cell with a high demand for ATP would minimize mitochondrial superoxide production, whereas a slowly dividing host cell might trigger the mitochondrial resting state, leading to maximal superoxide production (see Chance et al. 1979). Mitochondria would likely stimulate higher rates of mutation and sexual recombination in host cells with low cell division rates. (Blackstone 1995)

genetic recombination. In a finite population reproducing by asexual fission, sexual recombination coupled with selection allows deleterious mutations to be eliminated from the population (e.g., Bell 1988; Avise 1993). A higher mutation rate would select for a higher rate of sexual recombination (i.e., a lineage would be selected to initiate sexuality after fewer asexual generations). In this case, sexual recombination of the host would benefit the symbiont by providing novel genetic clones to infect (Fig. 4b). Once inhabiting a number of genetically novel hosts, protomitochondrial lineages would be buffered from environmental changes which could force whole clones of host cells to become extinct (see Williams 1975; Maynard Smith 1978).

Survival of the symbiotic association was likely contingent on rapid and profound evolutionary changes of the host in response to such manipulation. Many host cell lineages may have engaged in an evolutionary "arms race" with the protomitochondria (e.g., the host develops higher levels of antioxidants to suppress the protomitochondrion's signal, the protomitochondrion responds with higher levels of oxidants, etc.). Investment of energy in such an arms race would lead to compromised fitness at the level of the higher-level unit (e.g., Fig. 1) or perhaps directly to a disintegration of the association (e.g., the symbiont kills the host). By a variety of other mechanisms, other host lineages may have compromised their own replication rate in favor of that of the protomitochondrion (e.g., Fig. 1, Blackstone 1995). More successful host lineages would exhibit an adaptive response to the signaling mechanism of the endosymbiont, e.g., by dividing rapidly in response to an increased availability of ATP. Group selection acting on the protomitochondria would favor less destructive signaling mechanisms, e.g., the use of calcium fluxes. Because group selection of mitochondria will always be weaker than individual selection, selection on the higher-level unit (i.e., the host cell and the clonal mitochondrial population) would ultimately favor those lineages in which the ultimate control of the signaling mechanism was located in the nucleus. Thus the protomitochondrion would be turned into an organelle, unable to control its own energy allocation, replication rate, or intracellular signaling (Blackstone 1995).

The latter alternative has been shown to explain a number of the derived features of modern eukaryotic cells in their relationship to mitochondria (Blackstone 1995). These same considerations can also illuminate evolutionary considerations of the peroxisome. It is difficult to imagine that free-living protomitochondria did not have catalase; nevertheless this enzyme is not active in the mitochondrial compartment of modern eukaryotic cells (Fig. 2). Rather, catalase is the prototypical peroxisomal enzyme, and even cytoplasmic catalase is apparently derived from peroxisomes (e.g., Masters and Crane 1995). This arrangement is advantageous to the host cell because the properties of catalatic and peroxidatic reactions suggest that catalase is a more effective regulator of the concentration of intracellular H_2O_2 than the mitochondrial glutathione (Chance 1969; Chance et al. 1979). Mitochondria, even today, possess a small genome and are thus capable of heritable variation and evolution, while peroxisomes lack a genome and thus cannot be a unit of evolution. Locating catalase, the enzyme best able to regulate the concentration of intracellular H_2O_2, in the peroxisome effectively gives the nucleus greater control over H_2O_2 and thus control over signaling with this molecule as well.

The complex metabolic interplay between the mitochondrion and the peroxisome is also of note in this context. For instance, both mitochondria and peroxisomes are capable of β-oxidation of fatty acids, and there are enzymatic similarities and differences in this process in the two organelles (e.g., Masters and Crane 1995). Most remarkable are the substrate specificities of the two organelles. Short- and medium-chain fatty acids are not oxidized in the peroxisome but are preferentially oxidized in the mitochondria, and vice versa for long-chain fatty acids. Thus a long-chain fatty acid is first oxidized in the peroxisome and then completely oxidized in the mitochondrion, an ecological relationship which is in many ways reminiscent of the syntrophy commonly seen in communities of free-living microbes (Fenchel and Finlay 1995). Within the eukaryotic cell, however, this relationship suggests another nuance to the complex web of mechanisms designed to hold in check the evolution of selfish variant organelles (Blackstone 1995) by preventing either organelle from monopolizing the available substrate. It perhaps could be argued that there would be strong selection on the host cell to create the modern peroxisome even if there were no prospective symbiont readily available, and the similarities and differences between peroxisomal and mitochondrial genes (Borst 1989; Igual et al. 1992; Masters and Crane 1995) might be interpreted in light of this idea.

In any case, these mechanisms and those discussed in Blackstone (1995) allow the higher-level unit, the eukaryotic cell, to maintain a high fitness, largely uncompromised by selection on the lower-level units. Calcium signaling by mitochondria (and perhaps intracellular calcium signaling in eukaryotes in general) and the use of oxidants as signal molecules nevertheless persist as vestiges of the early history of the eukaryotic association, in a sense, a biochemical "balloon dance" (this in reference to the ritualization of the mating activities of emphidid flies, see Kessel 1955).

2.2.2
Single Cells to Multicellular Organisms

Evolution of multicellularity raises a number of the same selective issues as the evolution of the eukaryotic cell. As described by Buss (1987), the inability of metazoan cells to divide while flagellated suggests primitive multicellular morphologies in which ameboid cells are surrounded by flagellated cells. The ameboid cells carry out cell division and growth, whereas the flagellated cells provide locomotion. Further, it is the ameboid cells which reproduce these simple colonies. Selection at the level of the cell may favor "somatic cell parasites," that is, genetically variant cells which do not contribute to somatic duties but rather monopolize reproductive functions. Such genetically variant cells can potentially arise through copying errors during cell division. These variant cells will outcompete normal cells in the somatic environment (note the parallels to mammalian cancer cells). On the other hand, an organism composed of only "selfish" variants will quickly degenerate to the unicellular state, with the associated costs. Selection at the level of the cell (now the lower-level unit) will favor somatic cell parasites; selection at the level of the organism (the higher-level unit) will oppose these variants. Thus, for multicellularity to persist, mechanisms to attenuate this conflict in favor of the higher-level unit had to evolve (see Buss 1987).

As primitive multicellularity evolved in the ancestors of the metazoans, intercellular signaling likely became a key aspect of competition between cells. For instance, in the simple multicellular organism described above, a variant ameboid cell could enhance its relative replication rate simply by suppressing the replication of nearby cells and monopolizing the additional substrate thus made available. Mechanisms of intercellular signaling would have evolved in this context. From an evolutionary point of view, intercellular signaling is a mechanism which allows one cell to gain control of the intracellular signaling of another cell. Intracellular pathways involving calcium and oxidants are hypothesized to have originated from the units-of-evolution conflicts within eukaryotic cells; it is plausible that these same mechanisms could now be co-opted in this later units-of-evolution conflict between eukaryotic cells. The successful variant cell may have used a variety of messengers to trigger these intracellular pathways in neighboring cells, in a manner appropriate to cause the terminal differentiation of these cells. This in turn would have the dual benefits of automatically enhancing the variant cell's relative replication rate and of freeing additional metabolic resources with which this replication rate could be further enhanced. Simple metabolic gradients could be produced in this way.

As multicellular organisms became larger and more sophisticated, surface-to-volume constraints dictated the evolution of simple gastrovascular or circulatory systems. Such systems imply a centralized system for the distribution of food and other metabolites. In an organism with many totipotent or multipotent stem cells, allocation of resources can now become a key issue of conflict. If a cell or cell lineage does not receive enough substrate, selection will favor the cell lineage which initiates a series of morphogenetic changes to alter this situation, e.g., via angiogenesis or some related process. Selection will thus favor additional links between metabolic state and the proximate mechanisms of development.

The metabolic and oxidative gradients found in the development of modern metazoans may be a ritualized form of these sorts of interactions among cells and cell lineages early in the evolution of multicellularity. As more highly derived metazoans evolved, the role of these gradients may have been diminished and overwritten by other signaling mechanisms (cf. Newman 1992, 1994). In particular, the correlated evolution of the early sequestration of the germ line (Buss 1987) and other major changes in the genetic architecture of development (Davidson et al. 1995) may have relegated metabolic control to a secondary status in the embryonic development of protostomes and deuterostomes, although with certain exceptions, e.g., the vertebrate vascular system. Basal metazoans, however, may be expected to employ metabolic and oxidative mechanisms to a greater degree, in part because of their highly regulative development.

2.2.3 *Summary: The Logic of Metabolic Control*

There is a compelling evolutionary logic to the relationship between oxidative metabolism and development. In eukaryotes, much of the cell's oxidative metabolism is provided by organelles which likely had endosymbiotic origins. Metabolic signaling between these organelles and the nucleus may have evolved as a consequence of the conflicts between these evolutionary units. When multicellular organisms evolved and the evolutionary conflicts between the cell and the organism ensued, these intracellular signaling pathways may have been co-opted into intercellular signaling pathways. These latter pathways evolved to mediate cell-cell competition and remain as a vestige of this competition.

2.3 Summary

A modern mechanistic look at the relationship between metabolism and development suggests considerable support for what some may still consider an anachronistic viewpoint. While much work remains to be done, particularly in the invertebrates, an emerging consensus suggests that aspects of oxidative metabolism may play key roles in the signaling pathways which regulate development. Of particular importance in this regard are the use of intracellular oxidants and the direct signaling arising from the metabolic state of mitochondria. Since eukaryotic oxidative metabolism is primarily contained in the mitochondria and the peroxisomes, these principal organelles of eukaryotes are clearly implicated in the relationship between metabolism and development.

Considerations of the units-of-evolution conflicts surrounding the origins of mitochondria and peroxisomes suggest that natural selection may have favored the evolution of metabolic signaling between these organelles and the nucleus. Subsequently, to maintain the endosymbiosis, the evolution of nuclear control over these signaling mechanisms was necessary. Later in the history of life, these intracellular mechanisms may have been co-opted as intercellular signaling mechanisms during the evolution of multicellularity. These evolutionary considerations in fact predict the relationship between metabolism and development which modern studies of development now support.

3
Conclusions

Mechanisms that relate metabolism to development are presently being characterized in some vertebrates. These mechanisms may be found in all metazoans and may be particularly common in basal metazoans. Further, considerations of two key units-of-evolution transitions in the history of life suggest that there is an evolutionary basis for these mechanisms and predict the observed role of eukaryotic organelles in these mechanisms. Returning to the general systematic issues raised in the introduction, it is now possible to indicate the relationship between metazoan phylogeny and the mechanisms of development, and finally, to provide two predictions which can be tested by subsequent work.

3.1
Metazoan Phylogeny and the Mechanisms of Development

If metazoans are an independent experiment in multicellularity, many of the basic mechanisms of cell-cell interactions will have originated during the transition from unicellularity to multicellularity and will be shared derived characters. These mechanisms should be some subset of the mechanisms of development in modern metazoans. Here, aspects of these mechanisms which suggest a relationship between metabolism and development are focused on. The evolutionary history of eukaryotes, particularly the transition from simple to complex cells, may predispose all eukaryotes to use calcium fluxes and oxidative gradients as part of these developmental mechanisms. Use of oxidants, for instance, might be considered a "generic" mechanism in a sense parallel to that of Newman (1992, 1994). These generic mechanisms are expected to be used repeatedly in eukaryotic experiments in multicellularity. These features could be considered as shared primitive characters of all multicellular eukaryotes, although perhaps the concept of underlying synapomorphy is useful here (Sluys 1989). Nevertheless, the exact pathways by which these generic mechanisms act are expected to differ in each multicellular experiment.

Thus eukaryotic organelles may play a similar role in all eukaryotes, but the mechanisms by which these organelles carry out this role are expected to differ in specific detail among the eukaryotic groups which separately established multicellularity. The limited data that exist support this notion. For instance, O_2^- and H_2O_2 may act as intracellular second messengers in plants as well as animals (Sundaresan et al. 1995; Jabs et al. 1996), but the mechanisms for activating these messengers may differ (e.g., platelet-derived growth factor may affect intracellular H_2O_2 in animals, but not in plants, where, for instance, salicylic acid may bind to and inactivate catalase, thus affecting H_2O_2; see Sundaresan et al. 1995). Again, some role of mitochondria and peroxisomes in metabolic signaling may be a shared primitive character in all the multicellular eukaryotes, but the specific signaling mechanisms in which these organelles are involved may nevertheless be shared derived features for each multicellular group.

3.2
Toward a Predictive Theory of Evolution

To establish a predictive theory for the evolution of development, information from a variety of biological levels (e.g., biochemical, physiological, morphological, and ecological) must be integrated within a historical and phylogenetic framework. In this case the following predictions arise:

Hypothesis 1. A relationship between oxidative metabolism and multicellular development will exist in each group of multicellular organisms (animals, plants, fungi, etc.; anaerobic forms such as chytrids might be exceptions). "Generic" organelle-based mechanisms are expected to be present in each group.

Hypothesis 2. The specific details of the activation of these "generic" mechanisms of metabolism and multicellular development will be shared derived characters for each group of multicellular eukaryotes.

Acknowledgments. Comments on and criticism of the manuscript were provided by D. Bridge and L. Buss. The National Science Foundation (IBN-94-07049) and the Graduate School of Northern Illinois University provided support.

References

Adair TH, Gay WJ, Montani J (1990) Growth regulation of the vascular system: evidence for a metabolic hypothesis. Am J Physiol 259: R393–404

Alberts B, Bray D, Lewis J, Raff M, Roberts K, Watson JD (1983) The molecular biology of the cell. Garland, New York

Allbritton NL, Meyer T, Stryer L (1992) Range of messenger action of calcium ion and inositol 1,4,5-trisphosphate. Science 258: 1812–1815

Allen RG, Balin AK (1989) Oxidative influence on development and differentiation: an overview of a free radical theory of development. Free Rad Biol Med 6: 631–661

Ames B, Gold LS (1990a) Chemical carcinogenesis: too many rodent carcinogens. Proc Natl Acad Sci USA 87: 7772–7776

Ames B, Gold LS (1990b) Too many rodent carcinogens: mitogenesis increases mutagenesis. Science 249: 970–971

Avise JC (1993) The evolutionary biology of aging, sexual reproduction, and DNA repair. Evolution 47: 1293–1301

Backeljau T, Winnepenninckx B, De Bruyn L (1993) Cladistic analysis of metazoan relationships: a reappraisal. Cladistics 9: 167–181

Baltscheffsky H, Baltscheffsky M (1981) Mitochondrial ancestor models: *Paracoccus*, *Rhizobium*, and *Rhodospirillum*. In: Lee CP, Schatz G, Dallner G (eds) Mitochondria and microsomes. Addison-Wesley, New York, pp 519–540

Bandtlow CE, Schmidt MF, Hassinger TD, Schwab ME, Kater SB (1993) Role of intracellular calcium in NI-35-evoked collapse of neuronal growth cones. Science 259: 80–83

Bell G (1988) Sex and death in Protozoa. Cambridge University Press, Cambridge
Berridge MJ (1993) Inositol trisphosphate and calcium signaling. Nature 361: 315–325
Berridge MJ, Irvine RF (1989) Inositol phosphates and cell signaling. Nature 341: 197–205
Blackstone NW (1992) Heterochrony in hydractiniid hydroids: a hypothesis. In: Baskin A, Mittenthal J (eds) Principles of the organization of organisms. Addison-Wesley, Redwood City, California, pp 87–94
Blackstone NW (1995) A units-of-evolution perspective on the endosymbiont theory of the origin of the mitochondrion. Evolution 49: 785–796
Blackstone NW (in press) Physiological and biochemical aspects of dose-response for experimental heterochrony in colonial hydroids. J Evol Biol 10
Blackstone NW, Buss LW (1992) Treatment with 2, 4-dinitrophenol mimics ontogenetic and phylogenetic changes in a hydractiniid hydroid. Proc Natl Acad Sci USA 89: 4057–4061
Blackstone NW, Buss LW (1993) Experimental heterochrony in hydractiniid hydroids: why mechanisms matter. J Evol Biol 6: 307–327
Bonner JT (1996) Sixty years of biology. Princeton University Press, Princeton, New Jersey
Borst P (1989) Peroxisome biogenesis revisited. Biochim Biophys Acta 1008: 1–13
Brand MD, Murphy MP (1987) Control of electron flux through the respiratory chain in mitochondria and cells. Biol Rev 62: 141–193
Buss LW (1987) The evolution of individuality. Princeton University Press, Princeton, New Jersey
Carafoli E (1987) Intracellular calcium homeostasis. Annu Rev Biochem 56: 395–433
Cavalier-Smith T (1987) The simultaneous symbiotic origin of mitochondria, chloroplasts, and microbodies. Ann NY Acad Sci 503: 55–71
Chance B (1965) The energy-linked reaction of calcium with mitochondria. J Biol Chem 240: 2729–2748
Chance B (1969) The behavior of catalase and peroxidase in coupled reactions. Ann NY Acad Sci 168: 354–355
Chance B (1991) Optical method. Annu Rev Biophys Chem 20: 1–28
Chance B, Baltscheffsky H (1958) Respiratory enzymes in oxidative phosphorylation. J Biol Chem 233: 736–739
Chance B, Sies H, Boveris A (1979) Hydroperoxide metabolism in mammalian organs. Physiol Rev 59: 527–605
Child CM (1941) Patterns and problems in development. University of Chicago Press, Chicago
Child CM, Hyman LH (1919) The axial gradients in Hydrozoa. Biol Bull 36: 183–224
Coyne JA (1996) God in the details. Nature 383: 227–228
Davidson EH (1968) Gene activity in early development. Academic Press, New York
Davidson EH, Peterson KJ, Cameron RA (1995) Origin of bilaterian body plans: evolution of developmental regulatory mechanisms. Science 270: 1319–1325
de Duve C (1969) Evolution of the peroxisome. Ann NY Acad Sci 168: 369–381
Doolittle RF, Feng D-F, Tsang S, Cho G, Little E (1996) Determining divergence times of the major kingdoms of living organisms with a protein clock. Science 271: 470–477
Fanburg BL, Massaro DJ, Cerutti PA, Gail DB, Berberich MA (1992) Regulation of gene expression by O_2 tension. Am J Physiol 262: L235–L241
Fenchel T, Finlay BJ (1995) Ecology and evolution in anoxic worlds. Oxford University Press, Oxford
Floyd RA (1991) Oxidative damage to behavior during aging. Science 254: 1597
Galione A (1993) Cyclic ADP-ribose: a new way to control calcium. Science 259: 325–326
Gibbons GH, Dzau VJ (1996) Molecular therapies for vascular diseases. Science 272: 689–693
Gillham NW (1994) Organelle genes and genomes. Oxford University Press, Oxford
Gray MW (1989) Origin and evolution of mitochondrial DNA. Annu Rev Cell Biol 5: 25–50
Gray MW (1992) The endosymbiont hypothesis revisited. Int Rev Cytol 141: 233–335
Gray MW, Doolittle WF (1982) Has the endosymbiont hypothesis been proven? Microbiol Rev 46: 1–42
Halliwell B, Aruoma OI (1991) DNA damage by oxygen-derived species. FEBS Let 281: 9–19

Hansford RG (1985) Relation between mitochondrial calcium transport and control of energy metabolism. Rev Physiol Biochem Pharmacol 102: 1–72
Harman D (1981) The aging process. Proc Natl Acad Sci USA 78: 7124–7128
Heinle H, Tries S, Veigel C (1994) Is the influence of PO_2 or hydroperoxides on arterial tone mediated by the cellular thiol-disulfide system? Funktionsanalyse Biol Syst 22: 113–117
Holmes FL (1992) Manometers, tissue slices, and intermediary metabolism. In: Clarke AE, Fujimura JH (eds) The right tools for the job. Princeton University Press, Princeton, New Jersey, pp 151–171
Igual JC, González-Bosch C, Dopazo J, Pérez-Ortín JE (1992) Phylogenetic analysis of the thiolase family: implications for the evolutionary origin of peroxisomes. J Mol Evol 35: 147–155
Jabs T, Dietrich RA, Dangl JL (1996) Initiation of runaway cell death in an *Arabidopsis* mutant by extracellular superoxide. Science 273: 1853–1856
Jouaville LS, Ichas F, Holmuhamedov EL, Camacho P, Lechleiter JD (1995) Synchronization of calcium waves by mitochondrial substrates in *Xenopus laevis* oocytes. Nature 377: 438–441
Kessel EL (1955) Mating activities of balloon flies. Syst Zool 4: 97–104
Knoll AH (1992) The early evolution of eukaryotes: a geological perspective. Science 256: 622–627
Lawrence PA (1992) The making of a fly. Blackwell Scientific, Oxford
Lehninger AL (1964) The mitochondrion. WA Benjamin, New York
Lewontin RC (1970) The units of selection. Annu Rev Ecol Syst 1: 1–18
McCarroll R, Olsen GJ, Stahl YD, Woese CR, Sogin ML (1983) Nucleotide sequence of the *Dictyostelium discoideum* small-subunit ribosomal ribonucleic acid inferred from the gene sequence: evolutionary implications. Biochemistry 22: 5858–5868
Margulis L (1981) Symbiosis in cell evolution. WH Freeman, San Francisco
Masters C, Crane D (1995) The peroxisome: a vital organelle. Cambridge University Press, Cambridge
Maynard Smith J (1978) The evolution of sex. Cambridge University Press, London
Maynard Smith J, Szathmáry E (1995) The evolution of life's complexity. WH Freeman, San Francisco
Mitman G, Fausto-Sterling A (1992) Whatever happened to *Planaria*? C.M. Child and the physiology of inheritance. In: Clarke AE, Fujimura JH (eds) The right tools for the job. Princeton University Press, Princeton, New Jersey, pp 172–197
Morgan TH (1934) Embryology and genetics. Columbia University Press, New York
Morris PJ (1993) The developmental role of the extracellular matrix suggests a monophyletic origin of the kingdom Animalia. Evolution 47: 152–165
Morris PJ, CoBabe EA (1991) Cuvier meets Watson and Crick: the utility of molecules as classical homologies. Biol J Linn Soc 44: 307–324
Newman SA (1973) Reversible abolition of normal morphology in *Hydra*. Nat New Biol 244: 126–128
Newman SA (1992) Generic physical mechanisms of morphogenesis and pattern formation as determinants in the evolution of multicellular organization. J Biosci 17: 353–394
Newman SA (1994) Generic physical mechanisms of tissue morphogenesis: a common basis for development and evolution. J Evol Biol 7: 467–488
Nicholls DG, Akerman K (1982) Mitochondrial calcium transport. Biochim Biophys Acta 683: 57–88
Nijhout HF (1990) Metaphors and the role of genes in development. BioEssays 12: 441–446
Orr WC, Sohal RS (1994) Extension of life-span by overexpression of superoxide dismutase and catalase in *Drosophila melanogaster*. Science 263: 1128–1130
Richter C (1988) Do mitochondrial DNA fragments promote cancer and aging? FEBS Left 241: 1–5
Schilling WP, Elliot SJ (1992) Ca^{2+} signaling mechanisms of vascular endothelial cells and their role in oxidant-induced endothelial cell dysfunction. Am J Physiol 262: H1617–H1630
Schram FR (1991) Cladistic analysis of metazoan phyla and the placement of fossil problematica. In: Simonetta AM, Conway Morris S (eds) The early evolution of Metazoa and the significance of problematic taxa. Cambridge University Press, Cambridge, pp 35–46

Schram FR, Ellis WN (1994) Metazoan relationships: a rebuttal. Cladistics 10: 331–337
Shweiki D, Itin A, Soffer D, Keshet E (1992) Vascular endothelial growth factor induced by hypoxia may mediate hypoxia-initiated angiogenesis. Nature 359: 843–848
Siddall ME, Martin DS, Bridge D, Dresser SS, Cone DK (1995) The demise of a phylum of protists: phylogeny of myxozoa and other parasitic Cnidaria. J Parasitol 81: 961–967
Sluys R (1989) Rampant parallelism: an appraisal of the use of nonuniversal derived character states in phylogenetic reconstruction. Syst Zool 38: 350–370
Speksnijder JE, Terasaki M, Hage WJ, Jaffe LF, Sardet C (1993) Polarity and reorganization of the endoplasmic reticulum during fertilization and ooplasmic segregation in the ascidian egg. J Cell Biol 120: 1337–1346
Spenser DF, Schnare MN, Gray MW (1984) Pronounced structural similarities between small subunit ribosomal RNA genes of wheat mitochondria and *Escherichia coli*. Proc Natl Acad Sci USA 81: 493–497
Stadtman ER (1992) Protein oxidation and aging. Science 257: 1220–1224
Sundaresan M, Yu Z-X, Ferrans VJ, Irani K, Finkel T (1995) Requirements for generation of H_2O_2 for platelet-derived growth factor signal transduction. Science 270: 296–299
Szathmáry E, Maynard Smith J (1995) The major evolutionary transitions. Nature 374: 227–232
Thomas T, Thomas G, McLendon C, Sutton T, Mullan M (1996) β-Amyloid-mediated vasoactivity and vascular endothelial damage. Nature 380: 168–171
Tsien RW, Tsien RY (1990) Calcium channels, stores, and oscillations. Annu Rev Cell Biol 6: 715–760
Wainright PO, Hinkle G, Sogin ML, Stickel SK (1993) Monophyletic origins of the metazoa: an evolutionary link with Fungi. Science 260: 340–342
Wallace DC (1992) Mitochondrial genetics: a paradigm for aging and degenerative diseases? Science 256: 628–632
Williams GC (1975) Sex and evolution. Princeton University Press, Princeton, New Jersey
Williams GC (1992) Natural selection. Oxford University Press, Oxford
Wilmer P (1990) Invertebrate relationships-patterns in animal evolution. Cambridge University Press, Cambridge

Evolution of Animal Lectins

Jun Hirabayashi and Ken-ichi Kasai[1]

1 Introduction

Glycobiology is now becoming one of the most important fields in life science, although biological phenomena based on the specific recognition of sugar chains have attracted much less attention in comparison with those of nucleic acids and proteins in the past. Glycoscience is still in its infancy, and the biological significance of glycoconjugates is not readily understood. Too many questions remain unanswered, and no big discovery, such as the double helix structure of DNA, which changed our concept of life, has been made, nor have applications appeared such as gene engineering, which is significantly changing our way of life. However, in order to understand extremely complicated life systems, we now have to confront the biological phenomena based upon the specific recognition of sugar chains.

The recognition of sugar chains has completely different features from that of nucleic acids and proteins and is likely to be based on unknown principles. As informational biomolecules, sugar chains are peculiar. For example, why are sugar chains allowed to escape the direct control of the gene and be heterogeneous? Why are glycoproteins allowed to be modified by sugar chains heterogeneous in both size and structure, in spite of the fact that the protein parts are synthesized with the greatest care and strictness? Efforts to answer these fundamental questions will lead us to a variety of discoveries, and many unsolved problems in the life sciences will be solved and new principles of life discovered.

When unicellular organisms appeared on the primitive earth, nucleic acids and proteins would already have acquired their basic structures, and various important biological roles, especially informational ones, would have been assigned to them. On the other hand, sugar chains would have been less important as informational molecules, although they were indispensable molecules insofar as energy metabolism was concerned. However, when multicellular organisms were generated, communication between component cells would have become much more important. Then, the recognition of sugar chains would have become of greater significance, especially for multicellular

[1] Faculty of Pharmaceutical Sciences, Teikyo University, Sagamiko, Kanagawa 199-01, Japan

Progress in Molecular and Subcellular Biology, Vol. 19
W.E.G. Müller (Ed.)

organisms. The unique nature of sugar chains, e.g., heterogeneity, fuzziness, flexibility, and limited hereditability, etc., may have had certain advantages for such new purposes. Therefore, the necessity of investigating the specific recognition of sugar chains has now become extremely important. For this purpose, undoubtedly, understanding the counterparts of sugar chains, that is, lectins, is essential. Lectins are proteins that are distributed in almost all living organisms and specifically recognize the information presented by sugar chains of glycoconjugates. In this chapter, we discuss the evolution of animal lectins and, at the same time, the origin and evolution of the elementary saccharides that comprise informational sugar chains.

2 Brief Overview of Lectins

Since it is difficult to define lectins according to their biological roles because they are involved in too diverse biological phenomena, a definition based on the phenomenological point of view was once proposed: lectins are proteins having multiple sugar-binding sites and able to cause precipitation of polysaccharides and certain cells by making cross-links and networks between them (Goldstein et al. 1980). However, this definition turned out to be inappropriate because a number of monovalent saccharide-binding proteins that are structurally related to known lectins have been found. Therefore, at present, we must abandon the effort to establish a perfect definition and consider lectins to be sugar-binding proteins other than sugar-binding antibodies and enzymes acting on sugars. From the viewpoint of the transfer of biological information, lectins are therefore considered to be proteins that decipher glycocodes: each lectin recognizes a certain sugar structure and initiates a certain biological phenomenon in a given circumstance.

After the discovery of ricin in the caster bean by Stillmark in 1988, many plant lectins have been found. They have been used as very convenient research tools in the biosciences, e.g., for the determination of ABO blood types, detection and isolation of glycoproteins, analysis of the structure of saccharide chains, stimulation of cell division, etc., although their physiological functions have remained ambiguous, and it has not been easy to assign definite roles to them other than defense. However, research on animal lectins, which has become more and more extensive during the last two decades, has revealed the involvement of lectins in a variety of important phenomena that take place throughout the life of animals, e.g., fertilization, development, morphogenesis, differentiation, metastasis, apoptosis, etc.

A variety of possible roles attributed to lectins are listed in Table 1. Animals were also found to have lectins aimed at defense, examples being mammalian serum mannose-binding protein, which recognizes invading bacteria and destroys them by activating the complement system (Drickamer et al. 1986); sarcophaga lectin, which is induced in response to body injury (Takahashi et al. 1985); and macrophage lectin, which serves as a sensor molecule for invad-

Table 1. Possible functions of lectins

Function	Example
Defense	
Against infection	Activator of complement system (e.g., mannose-binding lectin)
	Sensor for invading organism (e.g., macrophage surface lectin)
	Immobilization of invading organism (e.g., invertebrate body fluid lectin)
Against Herbivora	Plant-toxic lectin (e.g., ricin)
	Inhibitor of insect amylase (e.g., legume lectin)
Cell-cell interaction	
Fertilization	Sperm lectin, ovum lectin
Migration	Homing of lymphocyte, migration of leucocyte, metastasis of malignant cell
Symbiosis	Adhesion of root nodule bacteria to legume root
Infection	Adhesion of bacteria to animal gut, infection of virus to host cell
Tissue organization	Condensation of mesenchymal cells (e.g., galectin)
Morphogenesis	
Cell-matrix interaction	Morphogenesis of dermis (e.g., galectin), guidance of axon filament (e.g., galectin)
Tissue organization	Material for matrix (e.g., proteoglycan core protein)
Adaptor	Cross-link between macromolecules
Regulation	
Enzyme activity	Thrombomodulin
Targeting of enzyme	Protease having C-type lectin domain
Cellular activity	Stimulation of mitosis, induction of apoptosis
Transportation	Carrier of lysosomal enzymes from the Golgi to the lysosome
Clearance	Removal of outdated serum proteins (e.g., hepatic lectin), scavenging of degraded cells
Weapon	Snake-venom lectin

ing microorganisms (Ii et al. 1990; Sato et al. 1992). There are many lectins involved mainly in cell-to-cell interaction and cell adhesion: both sperm and ova have lectins required for fertilization (Wassarmann 1987). The homing of lymphocytes and migration of leucocytes are performed by the guidance by a group of lectins named selectins (Lasky et al. 1989; Siegelman et al. 1989). Lectins are involved in the symbiosis of microorganisms with legume roots (Ho et al. 1986) and the infection of the animal gut by bacteria (Sharon 1987). The infection of microorganisms by viruses is also mediated by lectins. Mannose 6-phosphate-binding protein is an intracellular carrier molecule for the transfer of lysosomal enzymes from the Golgi apparatus to the lysosome (Kornfeld 1986). Mammalian hepatic lectin (asialoglycoprotein receptor) is the first animal lectin for which a definite biological role was attributed (Ashwell and Harford 1982). Hepatic lectin molecules are located on the surface of liver cells, bind serum glycoproteins that have lost terminal sialic acids of their sugar chains, and internalize them by endocytosis in order to remove outdated serum glycoproteins from the bloodstream. There are also lectins related to enzymes: some snake venom proteases have a lectin domain besides the active site (Takaya et al. 1992), and thrombomodulin is a regulator of the enzymes of blood clotting (Jackman et al. 1987). There are also lectins that seem to be involved in the construction of matrices, e.g., proteoglycan core protein (Doege et al. 1986; Sai et al. 1986) and galectins (Akimoto et al. 1992, 1995).

Up to now, two groups of animal lectins have been most extensively studied. One is the C-type lectin family (Drickamer 1988) and the other the galectin family (Barondes 1981, 1984; Barondes et al. 1994b; Caron et al. 1990; Drickamer and Taylor 1993; Hirabayashi 1993, 1994; Hirabayashi and Kasai 1993; Kasai 1990; Kasai and Hirabayashi 1996). Each family is characterized by a distinct amino acid sequence motif that builts the sugar-binding site (or CRD, carbohydrate-binding domain). The most distinct difference between these two families is the requirement for metal ions. The Ca ion is essential for C-type lectins to bind to sugar chains, but galectins do not need any metal ion. Besides a metal ion requirement, there are many other differences between these two families. Other lectin families, e.g., I-type lectin (Powell and Varki 1995) and A-type lectin (Kojima et al. 1992), have also been proposed. I-type lectin family members have the sequence motif of immunoglobulins, and an A-type lectin family was recently found to be the same as the anexin family. In this chapter, however, since it is still difficult to discuss the molecular evolutional aspects for these families, only those of C-type lectins and galectins are discussed. Comparison of these two families should be very interesting and extremely helpful in understanding the biological meaning of the recognition of sugar chains, because their natures are almost opposite in spite of the fact that they are both animal lectins.

3 Evolution of the C-Type Lectin Family

The name C-type lectin family was taken from the common requirement of its members for Ca ions for their sugar-binding ability (Drickamer 1988). A number of proteins having the characteristic motif, i.e., the carbohydrate-binding domain of C-type lectins (C-type CRD), have been found in a variety of animal species, although some proteins have not been confirmed to have sugar-binding ability, in spite of the presence of the typical C-type CRD. This is a very big family, and involves members with great diversity in terms of structure, sugar-binding specificity, and biological roles. Most of them are composed of several functional domains and are, therefore, multidomain proteins. Their biological roles are usually well defined as shown in Table 2. They can be divided into insoluble and soluble groups; and subsequently, each group is categorized into smaller groups in terms of protein architecture, biological function, distribution, and other properties (Table 2).

Table 2. Classified C-type lectin family

Group	Source	Specificity	Other domains	Reference
1. Receptor				
1.1 Type I receptor				
Selectins	Man, mouse	Sia + Fuc	TM, CR, EGF	Lasky et al. (1989); Siegelman et al. (1989)
Mannose receptor	Man	Man, Fuc	TM, FN-II, CYS	Lennartz et al. (1987)
Thrombomodulin	Man	?	TM, EGF	Jackman et al. (1987)
1.2 Type II receptor				
Hepatocyte receptor	Man, rat	Gal, GalNAc	TM	Drickamer et al. (1984)
	Chicken	GlcNAc	TM	Drickamer (1981)
Kupffer cell receptor	Rat	Gal, Fuc	TM	Hoyle and Hill (1988)
Macrophage receptor	Rat, mouse	Gal, GalNAc	TM	Ii et al. (1990)
HIV-gp 120 binding protein	Man	Fuc, Man	TM	Curtis et al. (1992)
2. Soluble protein				
2.1 Collagenous protein				
Mannose-binding protein	Liver (man, rat)	Man, Fuc > GlcNAc	CLG	Drickamer et al. (1986)
	Serum (rat)	Man, Fuc > GlcNAc	CLG	Drickamer et al. (1986)
Conglutinin	Cow	GlcNAC > Man > Fuc	CLG	Lee et al. (1991)
Pulmonary surfactant protein	Man, dog, rat	Fuc, Man, Gal	CLG	Benson et al. (1985); White et al. (1985)
2.2 Proteoglycan core protein				
Cartilage proteoglycan	Rat, chick	Gal, Fuc	GAG, HLA, Cys	Doege et al. (1986); Sai et al. (1986)
Fibroblast proteoglycan	Man	?	EGF, Cys	Krusius et al. (1987)
3. Snake venom protein				
Hemagglutinin	Rattlesnake	Gal		Hirabayashi et al. (1991)
Platelet agglutinin	Bothrop	?		Usami et al. (1993)
Coagulation factor inhibitor	Habu snake	?		Atoda et al. (1991)
Coagulation factor activator	Russell's viper	?	MP, DIS, CYS	Takeya et al. (1992)
Phospholipase inhibitor	Habu snake serum	?		Inoue et al. (1991)
4. Invertebrate body fluid lectin				
Galactose-specific lectin	Tunicate	Gal		Suzuki et al. (1990)
	Sea urchin	Gal		Giga et al. (1987)
	Fresh fly	Gal		Takahashi et al. (1985)
	Barnacle	Gal		Muramoto et al. (1986)
Regenectin	Cockroach	Sucrose		Kubo and Natori (1987)
5. Others				
Pancreatic stone protein	Man	?		DeCaro et al. (1987)
Regenerating islet protein	Man	?		Terazono et al. (1988)
Tetranectin	Man	?		Fuhlendorff et al. (1987)
Anti-freeze protein	Smelt, sea raven	?		Ewart et al. (1992), Ng et al. (1986)

3.1 Overview of the Members of the C-Type Lectin Family

3.1.1 *Receptor Proteins*

Type I Receptors. Type I receptors are membrane-integrated proteins whose N-terminal regions are exoplasmic and C-terminal regions, cytoplasmic. Their precursors may have a cleavable signal (leader) sequence. Though most receptor molecules take such an orientation, only a few examples have been found among C-type lectins. In these cases, C-type CRDs are located in the N-terminal region exposed on the cell surface. Thrombomodulin, known as a modulator of the blood coagulation system, is also categorized into this group (Jackman et al. 1987), although its sugar-binding ability has not been confirmed. It has extracellular EGF-like repeats (as many as six) as well as an N-terminal CRD-like structure.

Type II Receptors. A large percentage of membrane-integrated C-type lectins is categorized as type II receptors. They have an N-terminal cytoplasmic region, a hydrophobic transmembrane region, and a C-terminal exoplasmic region including CRD. They are synthesized without a cleavable signal sequence. Various well-known receptors, i.e., hepatic asialoglycoprotein receptor (Ashwell and Harford 1982; Leung et al. 1984), chicken hepatic lectin (Drickamer 1981), Kupffer cell receptor (Hoyle and Hill 1988), and macrophage asialoglycoprotein-binding protein (Ii et al. 1990) are typical type II receptors. They have well-defined functions: the removal of worn-out glycoproteins by endocytosis and phagocytosis. A placental mannose-specific protein known as an HIV gp120-binding protein has become a new member of this receptor subgroup (Curtis 1992).

3.1.2 *Soluble Proteins*

Collagen-Like Proteins (Collectins). This type is a chimeric protein composed of an N-terminal collagenous domain and a C-terminal CRD (Miyamura et al. 1994). The collagen-like regions of three such proteins combine to form a triple helix, resulting in the formation of a trimeric structure. Mannose-binding protein (Drickamer et al. 1986) and conglutinin (Lee et al. 1991) are supposed to activate the complement system by the classical pathway. The latter molecule has a significantly long collagenous domain, which is necessary for its function, because truncated forms of the mannose-binding protein lacking in this region failed to stimulate the complement system in vitro (Schweinle et al. 1993). A major component of pulmonary surfactant, which potentiates the surface tension-reducing properties of surfactant lipids, is the first "lectin-related" protein revealed by cDNA cloning (Benson et al. 1985; White et al. 1985; Haagsman et al. 1987).

Proteoglycan Core Proteins. Proteoglycan is a very large glycoprotein consisting of various domains, e.g., a large number of attachment sites for glycosaminoglycans, hyaluronic acid, etc. cDNA sequences for proteoglycans of rat and chicken cartilage and human fibroblast have been reported (Doege et al. 1986; Sai et al. 1986; Krusius et al. 1987). CRDs are located in their C-terminal regions. The segment encompassing the C-terminal 130 residues of rat cartilage proteoglycan showed fucose- and galactose-specific lectin activity (Halberg et al. 1988). The biological significance, however, of the involvement of C-type CRDs in the proteoglycans remains unclear.

Snake-Venom Lectins and Lectin-Related Proteins. Crotalid and viperid snakes have a set of characteristic venomous proteins, including various hydrolytic enzymes, that cause complex local injury. The venoms also contain galactose-specific lectins that show strong hemagglutinating ability. The primary structure of a venom lectin from the rattle snake *Crotalus atrox* clearly showed that this lectin belongs to the C-type lectin family (Hirabayashi et al. 1991). These lectins also stimulate platelet aggregation in a sugar- and calcium-dependent manner.

As lectin-related molecules, an inhibitor of coagulation factor X and an activator of coagulation factor IX/X have been isolated from Habu snake and viper venoms, respectively (Atoda et al. 1991). The latter is a two-chain-composed multidomain protein. The heavy chain has metalloprotease activity, whereas the light chain consists of only a CRD. Besides venom, Habu snake serum contains a kind of phospholipase inhibitor that turned out to contain the C-type lectin CRD (Inoue et al. 1991).

Invertebrate Body Fluid Lectins. C-type lectins have been found in the body fluids of various invertebrate species, e.g., tunicate (Suzuki et al. 1990), sea urchin (Giga et al. 1987), fresh fly (Takahashi et al. 1985), cockroach (Kubo and Natori 1987), and barnacle (Muramoto and Kamiya 1986). Most of them are galactose-specific lectins, although a cockroach hemolymph protein shows specific affinity for the lipopolysaccharide of *E. coli.* Usually, the structure of their monomer unit is simple because they are essentially composed of only one CRD. However, they usually form polymeric structures of high molecular weight under nondenaturing conditions. They are assumed to be substitutes for immunoglobulins and involved in the defense against bacterial infection. The lectin of sarcophaga also plays an important role during metamorphosis.

Others. Pancreatic stone proteins, being present in pancreatic juice and inhibiting the growth of calcium carbonate crystals (De Caro et al. 1987) and Langerhans islet regeneration-specific protein (Terazono et al. 1988), were recently proved to be the same protein. Tetranectin, a regulator of the blood coagulation system (Fuhlendorff et al. 1987), and a type II antifreeze protein found in smelt and sea-raven sera are also soluble, C-type lectin-related proteins (Ng et al. 1986; Ewart et al. 1992). All these molecules have a simple structure consisting of only one CRD, although their sugar-binding properties have not been confirmed.

3.2 Biochemical Properties of C-Type Lectins

3.2.1 *General Properties*

All of the C-type lectins and structurally related proteins, including receptors, have primary structures typical to extracellular proteins or membrane-anchored proteins. Their nascent peptide has either a cleavable N-terminal signal sequence or a hydrophobic sequence, required for insertion into the membrane. Many C-type lectins are glycosylated, and all of them have stable intramolecular disulfide bonds. The presence of two highly conserved disulfide bonds is one of the best proofs of their assignment to the C-type lectin class.

C-type lectins require calcium ions on the order of 1 mM concentration to express their activity. Since calcium binding is reversible, this property has often been utilized for the purification of these lectins by affinity chromatography. This should also be of biological importance, because lectin activity could be regulated by a change in the calcium ion concentration. A recent study on the biosynthesis of the hepatic asialoglycoprotein receptor suggested that the calcium ion is also required for the translocation of this receptor lectin from the endoplasmic reticulum to the Golgi apparatus (Lodish et al. 1992).

3.2.2 *Binding Specificity*

Sugar-binding specificities of C-type lectins are diverse; there are lectins specific for mannose, N-acetylglucosamine, galactose, fucose and sialic acid. However, systematic studies of the sugar-binding properties of three C-type lectins, each having a different specificity (N-acetylglucosamine-specific chicken hepatic lectin, galactose-specific rat hepatic lectin, and rat serum mannose-binding protein), showed that they have several common properties, that is: (1) they recognize only non-reducing terminal monosaccharide moieties; (2) 3- and 4-, but not 2-, hydroxyl groups of the target monosaccharide are important for recognition; and (3) they show extremely high affinity for branched sugar chains (clustering effect) (Lee et al. 1984, 1989a,b).

These results suggest that the global conformation of the CRD has been preserved even though its specificities were changed. These results also explain well why most mannose-specific lectins have an affinity for N-acetylglucosamine besides mannose, because the substituting group at the C-2 position is not critical, and why hydroxyl groups of C-3 and C-4 positions are often used for the formation of glycosidic bonds in sugar chains. The clustering effect seems the most important characteristic of some C-type lectins, such as asialoglycoprotein receptor. In fact, they show an extremely enhanced affinity for bi- and triantenary sugar chains. This suggests that these C-type lectins adopt highly organized oligomeric structures.

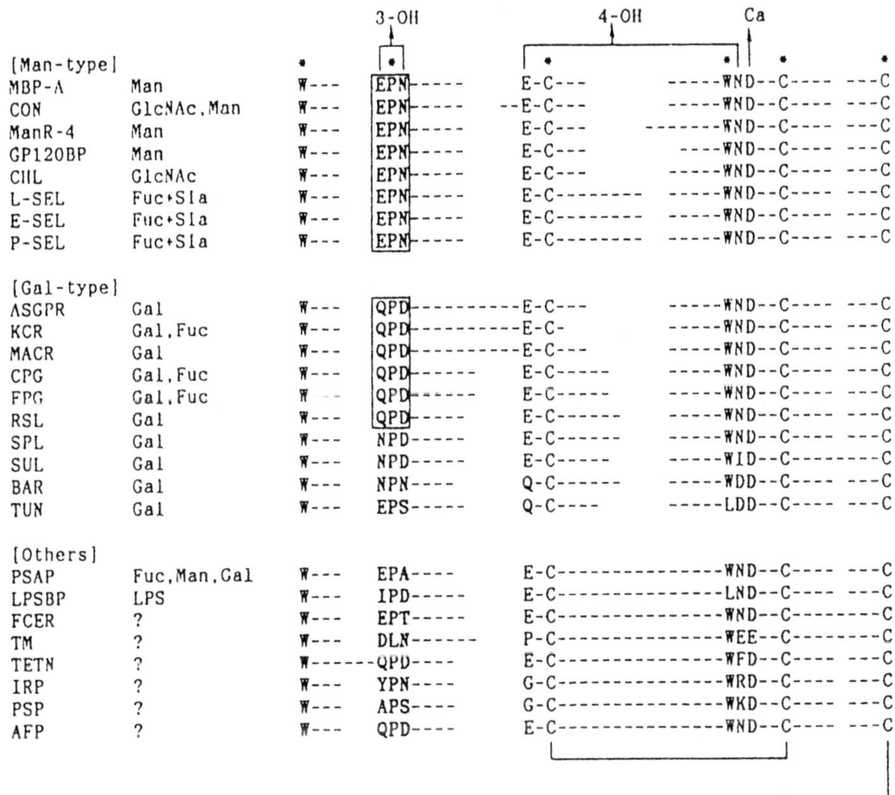

Fig. 1. Sequence motifs defining sugar-binding specificities of C-type lectins. Only relevant residues are shown (others shown by *dashed lines*). *Asterisks* denote structurally invariant residues. *Blanks* represent deletions for maximum homology. Motifs, *EPN*, specific for mannose-binding lectins, and *QPD*, specific for galactose-binding lectins are *boxed*. Abbreviations: *MBP-A* mannose-binding protein-A; *CON* conglutinin; *ManR-4* macrophage mannose receptor CRD-4; *GP120BP* placental HIV gp12-binding protein; *CHL* chicken hepatic lectin; *L-*, *E-*, *P-SEL* L-, E-, P-selectins, respectively; *ASGPR* hepatic asialoglycoprotein receptor; *KCR* Kupffer cell receptor; *MACR* macrophage receptor; *CPG* cartilage proteoglycan core protein; *FPG* fibroblast proteoglycan core protein; *RSL* rattle snake venom lectin; *SUL* sea urchin lectin; *BAR* barnacle lectin; *TUN* tunicate lectin; *PSAP* major component of pulmonary surfactant apoprotein; *LPSBP* cockroach lipopolysaccharide-binding protein; *FCER* low-affinity IgE receptor (CD23); TM thrombomodulin; *TETN* tetranectin; *IRP* islet regenerating protein; *PSP* pancreatic stone protein; *SPL* sarcophaga lectin; and *AFP* smelt anti-freeze protein

3.2.3
Structure of the Sugar-Binding Site

Figure 1 shows the amino-acid sequences of the essential parts of the CRD of various C-type lectins. It is evident that a number of C-type lectins having specificity for mannose, N-acetylglucosamine, and L-fucose have the Glu-Pro-Asn sequence, whereas those having specificity for galactose tend to have to Gln-Pro-Asp sequence. These results suggest that the amino-acid side chains on both sides of Pro186 (numbered according to the sequence of rat mannose-binding protein) are critical determinants of ligand binding.

The three-dimensional structure of one of the C-type lectins, the rat mannose-binding protein, complexed with an Asn-linked biantenary sugar chain, was solved (Drickamer 1992; Weis et al. 1992). It was shown that: (1) amino-acid residues involved in the sugar binding were located in the C-terminal region encompassing 25 residues; (2) two "acid-amide" pairs, Glu185/Asn 187 and Glu193/Asn205, recognized the 3- and 4-hydroxyl groups of mannose, respectively, by cooperative hydrogen bonding; and (3) these hydrogen bonds were strictly coordinated by the presence of one calcium ion. One of the pairs, Glu185/Asn187, was found to be critical for the

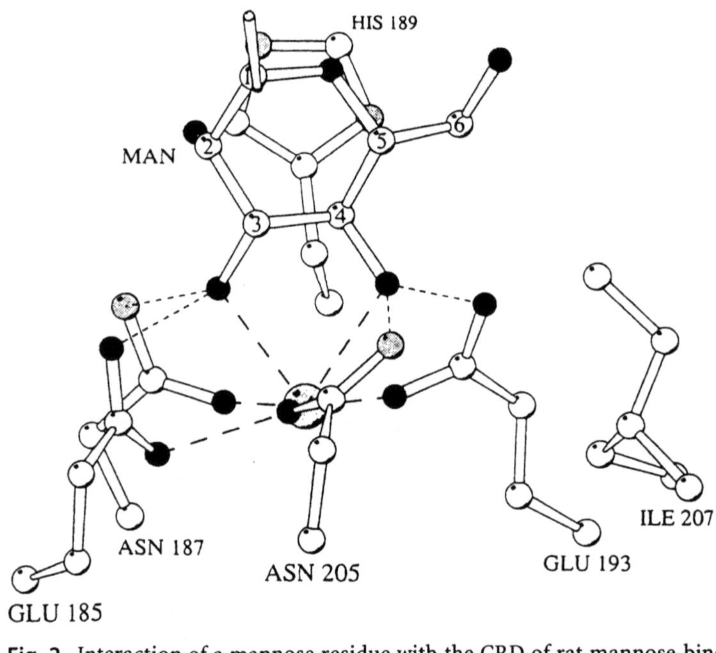

Fig. 2. Interaction of a mannose residue with the CRD of rat mannose-binding protein. *Dotted lines* show hydrogen bonds and the *large shaded sphere* is Ca ion. (Iobst et al. 1994)

sugar-specificity determination when various saccharide molecules other than mannose were superimposed on a computer screen. Glucose and N-acetylglucosamine have the same orientation as mannose in terms of 3- and 4-hydroxyl groups (Fig. 2). Therefore, they can be accommodated in the binding site if there is no steric hindrance around the C-2 hydroxyl group. L-fucose is also accommodated because its 2- and 3-hydroxyl groups have the same orientation as the 3 and 4 positions of mannose if its ring is inverted. Galactose could not be accommodated because the axial C-4 hydroxyl group caused steric hindrance.

Substitution by recombinant DNA technology of the sequence Glu185-Pro186-Asn187 of rat mannose-binding protein, the characteristic motif of mannose-specific lectins, with Gln185-Pro186-Asp187, that of galactose-specific lectins, resulted in the change of sugar specificity, i.e., the mannose-binding protein became galactose-specific, although this mutant protein retained half of its original affinity for mannose (Drickamer 1992).

3.3 Molecular Evolution of C-Type Lectins

Members of the C-type lectin family are distributed very widely. They have been found in both protostomes and deuterostomes, although none has been found in sponges. There should have been a common ancestor of all C-type lectins before the divergence of protostomes and deuterostomes. It is not easy to discuss the molecular evolution of the C-type lectin family on the basis of their entire structure, because they have diverged into several subfamilies in which both protein architecture and biological function are too diversified. Therefore, the discussion should be limited to the evolution of only the CRD region. Drickamer (1993) has written a review on the evolution of C-type lectins and proposed a dendrogram based on the comparison of CRD regions of about 50 C-type lectins (Fig. 3). The similarities of CRDs between various combinations were found to be 20–60%. It is obvious that sequence similarities were higher within each group classified according to their molecular organizations and functions. This suggests that all proteins belonging to a single group descend from a common ancestor that already had the characteristics presently observable.

The genes for C-type CRD regions may have combined with genes for various functional motifs of various proteins, and, at the same time, mutations within CRD regions would have occurred, resulting in the change in the binding specificity. Drickamer (1993) suggested that a putative primitive binding domain capable of binding a variety of molecules and requiring Ca ion for stabilization of its shape evolved into a sugar-binding domain and that this ancestral CRD was specific for mannose and N-acetylglucosamine, because

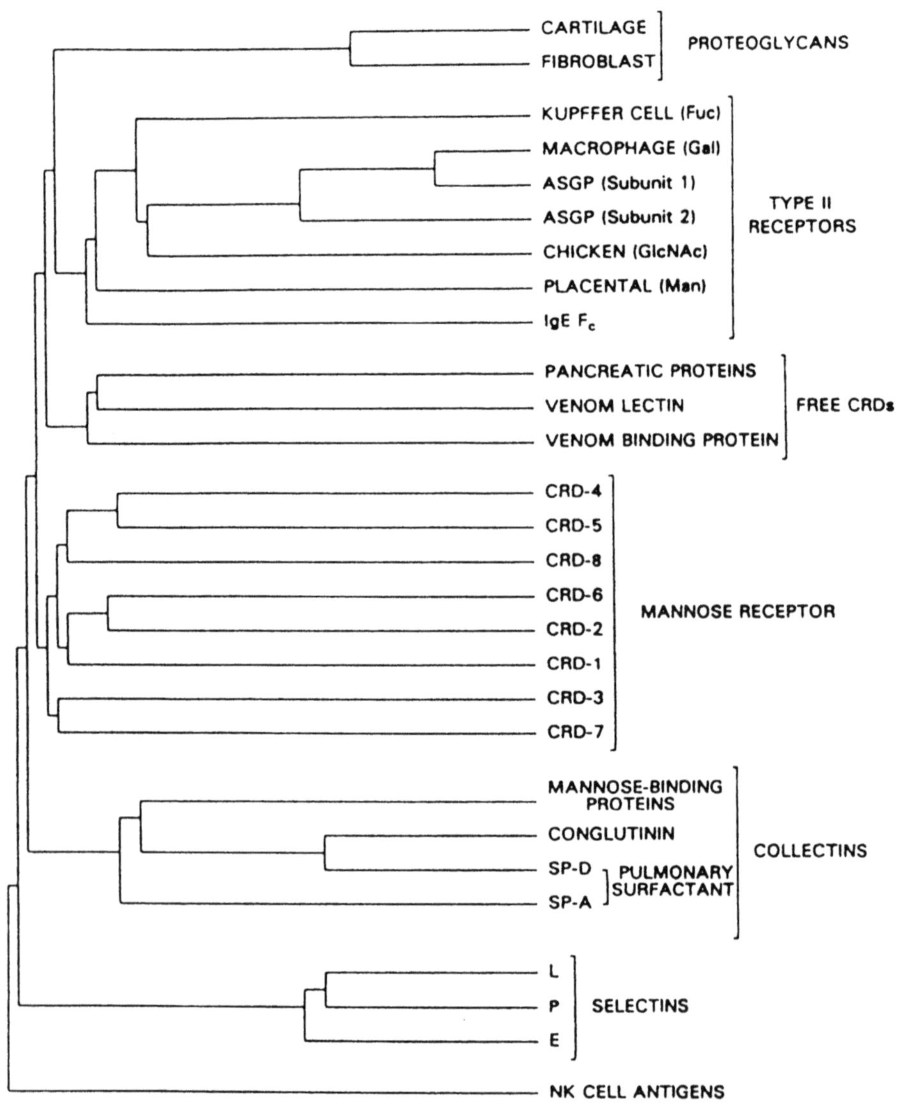

Fig. 3. Dendrogram proposed by Drickamer (1993) on the basis of sequence similarity between various C-type CRDs. Mannose receptor contains 8 CRDs. The horizontal scale cannot be used to infer specific times of evolutionary divergence

each group has a common mannose-binding precursor but not a common galactose-binding progenitor. CRDs specific for galactose in each group seem to have evolved independently from these precursors, and a change in the Glu-Pro-Asn motif specific for mannose and N-acetylglucosamine (and L-fucose) to the Gln-Pro-Asp sequence occurred. Therefore, convergent evolution seems to have occurred during these processes to achieve similar binding specificities (Drickamer 1993).

4 Evolution of Galectins

4.1 Galectin: A New Family Name for a Group of Animal Lectins

4.1.1 *The Early Period of Galectin Research*

Galectin is a new family name created for a group of animal lectins which are very different from C-type lectins. Since the first report on the presence of a galactoside-specific lectin in electric eel (Levi and Teichberg 1981), animal lectins having common properties, that is, specific for β-galactosides and not dependent on metal ion, have been found in various vertebrates (e.g., frog, chicken, mouse, human, etc.). They have been given different names depending on the researchers (e.g., electrolectin, soluble lectins, galaptins, S-Lac lectins, β-galactoside-binding lectins, S-type lectins, etc.). Because they turned out to have sequences homologous to each other, and the number was growing, the necessity for a generic family name became very great in order to avoid inconvenience and confusion. Then, agreement was reached among the involved researchers to use the new term "galectin" (Barondes et al. 1994a). It was also agreed to give a serial number to mammalian galectins in the order of discovery (Table 3).

In vertebrates, galectins were found in various tissues and cells, e.g., skin, muscle, brain, intestine, liver, kidney, placenta, cultured fibroblasts, established cell lines such as HL60, and tumor cells. A number of observations have indicated their possible involvement in a variety of important phenomena occurring in multicellular animals, such as development, differentiation, morphogenesis, immunity, apoptosis (Perillo et al. 1995), etc. Metastasis of malignant cells was also reported to be related to the presence of galectins on the cell surface (Raz and Lotan 1987). Various cellular activities were found to be triggerd by galectins (Wells and Malluci 1991; Yamaoka et al. 1991). The biological functions of galectins appear to vary from location to location and from time to time. Too many cases in which the involvement of galectins are suggested make it almost impossible to define their principal mission. From the viewpoint of mechanism, various cellular processes, such as communication, adhesion, transportation, guidance, modulation, signal transduction, etc., seem to be supported by the specific recognition of glycoconjugates by galectins.

Galectins found in the early period were noncovalent homodimers of a subunit of about 14 kDa, which constitutes the minimum structural element (prototype) of the galectin family. Galectins of this type usually show

Table 3. Galectins classified according to molecular architecture

Type	Animal class	Species	Example	Reference
Proto	Mammals	Man, mouse, rat, cow, *etc.*	Galectin-1	Hirabayashi and Kasai (1984, 1988); Clerch et al. (1988); Abbott et al. (1989); Merkle et al. (1989); Wilson et al. (1989)
		Man	Galectin-2	Gitt et al. (1992)
		Rat, mouse	Galectin-5	Gitt et al. (1995)
			Galectin-7	Madsen et al. (1995); Magnaldo et al. (1995)
	Birds	Chicken, quail	14 kDa	Ohyama et al. (1986); Hirabayashi et al. (1987)
			16 kDa	Sakakura et al. (1990)
	Amphibia	Frog (*Rana, Xenopus*)	14 kDa, 16 kDa	Marschal et al. (1993); Ozeki et al. (1991)
	Fish	Electric eel, conger eel	14 kDa, 16 kDa	Paroutaud et al. (1987); Muramoto and Kamiya (1986)
	Nematodes	*C. elegans*	16 kDa	Hirabayashi et al. (1996)
	Sponges	*Geodia cydnium*	13–16 kDa	Pfeifer et al. (1993)
Chimera	Mammals	Man, rat, mouse	Galectin-3	Jia and Wang (1988)
	Birds	Chick	Galectin-3	Nurminskaya and Linsenmayer (1996)
Tandem repeat	Mammals	Rat, mouse	Galectin-4	Oda et al. (1993)
			Galectin-6	Gitt et al. (1995)
			Galectin-8	Hadari et al. (1995)
		Man, mouse, rat	Galectin-9	Sahin et al. (1995)
	Nematodes	*C. elegans*	32 kDa	Hirabayashi et al. (1992)
		Onchocerca volvolus	32 kDa	Klion and Donelson (1994)

hemagglutinating activity. Determination of the complete primary structure of chicken 14-kDa galectin (C-14; Hirabayashi et al. 1987) and cloning of its cDNA (Ohyama et al. 1986) triggered extensive structural studies on galectins from various sources, and the primary structures of a number of mammalian galectins (e.g., galectin-1 from man; Hirabayashi and Kasai 1988), cow (Abbott et al. 1989), rat (Clerch et al. 1988; Leffler et al. 1989), mouse (Wilson et al. 1989)), and chicken 16 kDa galectin (C-16; Sakakura et al. 1990) were elucidated successively. Presence of the second human galectin (now designated galectin-2) was also reported (Gitt and Barondes 1986). Gene engineering has also been applied to galectin research (Hirabayashi and Kasai 1989, 1991; Abbott and Feizi 1991), and various genetically engineered galectins have been produced (Abbott et al. 1991; Hirabayashi and Kasai 1991, 1994). Alignment of the revealed sequences indicated that the central parts of the galectin molecule are mostly conserved and presumably correspond to the sugar-binding domain (CRD; Fig. 4). Characteristic residues that compose the galectin CRD are shown in large capital letters in Fig. 5. Similarities between mammalian galectins found during this early period of research were significant, and they are now grouped in the galectin-1 category.

4.1.2
Expansion of the Galectin Family

On the other hand, a larger galactose-binding protein of about 35 kDa (CBP35) was found in 3T3 cells. Determination of its primary structure revealed that it contained the galectin CRD motif in the C-terminal half-region (Fig. 5; Jia and Wang 1988). This was the first indication of the existence of the galectin family. The sequences of the N-terminal half, however, had no homology to galectins. This protein, recently named galectin-3, proved to be the same as the proteins that had been called either IgE-binding protein of certain leucocytes (Albrandt et al. 1987) or Mac-2 (a macrophage-specific cell-surface antigen; Cherayil et al. 1989).

Up to now, discoveries of proteins having the galectin CRD but distinct from known galectins have been reported continuously, especially in the case of mammals (Table 3). The numbered mammalian galectins so far have reached as many as 9, though it is not yet clear why there should be so many galectins, and it is not possible to assign a specific role to each member.

Though for a long time galectins were found only in vertebrate species, a galectin of 32 kDa was found in the nematode *Caenorhabditis elegans* (Hirabayashi et al. 1992; Arata et al. 1996). This was the first discovery not only of a nematode *lectin* but also of a new type of the galectin family. The molecular architecture of the 32-kDa molecule was unique, because two CRD motifs were found to be tandemly repeated in a single polypeptide, suggesting that it is a divalent monomer (Fig. 5). The existence of a mammalian galectin having a similar protein architecture was also reported recently (named galectin-4; Oda et al. 1993). In addition, a prototype *C. elegans* galectin having a molecular

```
                                                                                              # # #      # #
Gal-1 (human)                         MACGLVASNLNLKPGEC-LRVRGEVAPDAKS-FVLNLGKDSN-----NLCLHFNPRFNAHGDANTIVCNS
Gal-2 (human)                         MTGELEVKNMDMKPGST-LKITGSIADGTDG-FVINLGQGTD-----KLNLHFNPRFSES----TIVCNS
Gal-3 (human)                     YGAPAGPLIVPYNLPLPGGVVPRML-ITILGTVKPNANR-IALDFQRGND------VAFHFNPRFNEN-NRRVIVCNT
Gal-4I (rat)               MAYVPAPGYQPTYNPTLPYKRPIPGGLSVGMS-IYTQGIAKDNMRR-FHVNFAVGQD--EGADIAFHFNPRFDGW---DKVVFNT
Gal-4II                 ...MNSLPVMAGPPIFNPPVPYVGTLQGGLTARRT-IIIKGYVLPTAKN-LIINFKVGST----GDIAFHMNPRI-----GDCVVRNS
Gal-5 (rat)                  MSSFSTQTPYPNLAVPFFTSIPNGLYPSKS-IVISGVVLSDAKRRFQINLRCGGD------IAFHLNPRFDEN----AVVRNT
Gal-7 (human)                         MSNVPHKSSLPEGIRPGTV-LRIRGLVPPNASR-FHVNLLCGEF--QGSDAALHFNPRLDTS----EVVFNS
Gal-8I (rat)                 MLSLSNLQNIIYNPTIPYVSTITEQLKPGSL-IVIRGHVPKDSER-FQVDFQHGNSLKPRADVAFHFNPRFKRS---NCIVCNT
Gal-8II                                  ...PFEARLNASMGPGRT-VVVKGEVNTNATS-FNVDLVAGRS----RDIALHLNPRLNVK----AFVRNS
CLC-LPLase (human)                  MSLLPVPYTEAASLSTGSTVT-IKGRPLVCFLNEP-YLQVDFHTEM-KEESDIVFHFQVCFGRR-----VVMNS
Chick-14K                             MSCQGPVCTNLGLKPGQR-LTVKGIIAPNAKS-FVMNLGKDST-----HLGLHFNPRFDAHGDVNLIVCNS
Chick-16K                              MEQGLVVTQLDVQPGEC-VKVKGKILSDAKG-FSVNVGKDSS-----TLMLHFNPRFDCHGDVNTVVCNS
Frog                                   MDMEPDVRITNLNLHKGHR-VEVRGRIAKGTNR-FAVDLGTDSR-----NLICHCNPRFEYSVDKNTIVLNS
Conger eel                               SGGLQVKNFDFTVGKF-LTVGGFINNSPQR-FSVNVGESMN-----SLSLHLDHRFNYGADQNTIVMNS
Nematode-16K              MIGGGIGISFCNEFFNPQTPVNIPVQGFSNGAR-LRLVLLPTSADSR-FHINLRTPDD------IVLHFNARFDE----GAVVNNS
Nematode-32KI                  MSAEEPKSYPVPYRSVLQEKFEPGQT-LIVKGSTIDESQR-FTINLHSKTADFSGNDVPLHVSVRFDE----GKIVLNS
Nematode-32KII                         KYYPVPYESGLANGLPVGKS-LLVFTTVEKKAKR-FHVNLLRKN-----GDISFHFNPRFDE----KHVIRNS
Sponge-I                ...MQEINYDPYDEDDRYNSAFAQVFGDLKLTVPSLTVGQTIEVLYVNPPTGALSINLVTAD-DDVALHFNPRYSSTGG--YLVLNT
Sponge-II               ...MME--YGTDDGVEQYNSASAQVFGDLKVTVPSLTVGQTVEAVFINPPTGVLSLNLVTGNGNDVALHFNPRHASTGG--LLVLNS
```

```
                      #  # #
Gal-1 (human)      --KDGGAWGTEQREAV--FPFQPGSVAEVCITFDQANLT---VKLPDGYEFKFP-NRL-NLEAINYMAADGDFKIKCVAFD.
Gal-2 (human)      --LDGSNWGQEQREDH--LCFSPGSEVKFTVTFESDKFK---VKLPDGHELTFP-NRL-GHSHLSYLSIRGGFNMSSFKLKE.
Gal-3 (human)      --KLDNNWGREERQSV--FPFESGKPFKIQVLVEPDHFK---VAVNDAHLLQYN-HRVKKLNEISKLGISGDIDLTSASYTMI.
Gal-4I (rat)       --MQSGQWGKEEKKKS--MPFQKGHHFELVFMVMSEHYK---VVVNGTPFYEYG-HRL-PLQMVTHLQVDGDLELQSINFL...
Gal-4II            --YMNGSWGSEERKIP-YNPFGAGQFFDLSIRCGTDRFK---VFANGQHLFDFS-HRFQAFQRVDMLEIKGDITLSYVQI.
Gal-5 (rat)        --QINNSWGPEERSLPGSMPFSRGQRFSVWILCEGHCFK---VAV-DGQHICEYSHRLMNLPDINTLEVAGDIQLTHVET.
Gal-7 (human)      --KEQGSWGREERGPG--VPFQRGQPFEVLIIASDDGFK---AVVGDAQYHHFR-HRL-PLARVRLVEVGGDVQLDSVRIF.
Gal-8I (human)     --LTNEKWGWEEITHD--MPFRKEKSFEIVIMVLKNKFH---VAVNGKHILLYA-HRI-NPEKIDTLGIFGKVNIHSIGFRFS...
Gal-8II            --FLQDAWGEEERNIT--CFPFSSGMYFEMIIYCDVREF--KVAVNGVHSLEYK-HRFKDLSSIDTLAVDGDIRLLDVRSW.
CLC-LPLase         --REYGAWKQQVESKN--MPFQDGQEFELSISVLPDKYQ---VMVNGQSSYTFD-HRI-KPEAVKMVQVWRDISLTKFNVSYLKR.
Chick-14K          --KKMEEWGTEQRETV--FPFQKGAPIEITFSINPSDLT---VHLP-GHQFSFP-NRL-GLSVFDYFDTHGDFTLRSVSWE.
Chick-16K          --KEDGTWGEEDRKAD--FPFQQGDKVEICISFDAAEVK---VKVP-EVEFEFP-NRL-GMEKIQYLAVEGDFKVKAIKFS.
Frog               --KQNDVWDIEKKETA--FPF------KMLIFDFEDCIT---VHLPDGKEIPFT-CRF-PIEVINYLALNNIELISISVH.
Conger eel         TLKGDNGWETEQRSTN--FTLSAGQYFEITLSYDINKFY---IDILDGPNLEFP-NRY-SKEFLPFLSLAGDARLTLVKE.
Nematode-16K       --TSGGGWQSEDRHA---NPFQQNKIYTLEFVSNGGIIS---IFVNGAHFADFV-ERTPSHGVHLIEIEGGVHVHSAHVSH.
Nematode-32KI      --FSNGEWGKEERKS---NPIKKGDSFDIRIRAHDDRFQ---IIVDHKEFKDYE-HRL-PLSSISHLSIDGDLYLNHVHWGG...
Nematode-32KII     --LAANEWGNEEREGK--NPFEKGVGFDLVIQNEEYAFQ---VFVNGERYISFA-HRA-DPHDIAGLQISGDIELSGIQIQ.
Sponge-I           --LLNGNWQTEVHPTG--FPFPANNVKTRVLVSITVQEKDFLLQVNGIDITTFS-YHP-GLSYDKVRHITCKGLEHAVLEAIGVTY.
Sponge-II          --LMNGVWADEVRPAG--YPFPFYNVQTRTTVRITVKEDGFLITVNGINITTFM-YRP-GLGYETVRHITWKAVPEAILESIQITY.
```

Fig. 4. Sequence alignment showing amino acid sequence identities to human galectin-1 (Hirabayashi and Kasai 1988). Residues identical to those of human galectin-1 are *shaded*. Deletions introduced for maximum homology are denoted by (-). Residues proved to interact with the bound lactose molecule by X-ray crystallography are denoted by (#). (Lobsanov et al. 1993)

weight of approximately 16 kDa was also found (Hirabayashi et al. 1996). Therefore, the nematode is likely to be equipped with a set of galectins similar to those in vertebrates, although a chimera-type galectin has not yet been found. Soon after the discovery of the galectin in *C. elegans*, the existence of galectins in the marine sponge *Geodia cydonium* was reported (Pfeifer et al. 1993). This finding has a significant meaning for galectin research, because the phylogenically oldest metazoan phylum (Porifera) already had galectins. Therefore, galectins must have appeared on the earth from the beginning of

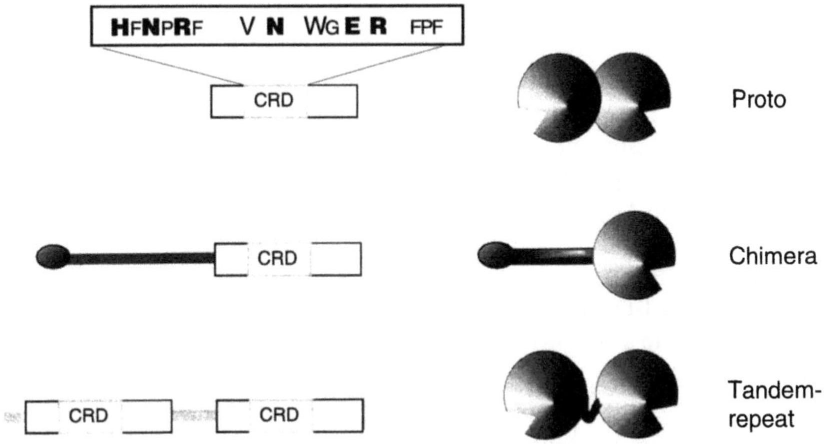

Fig. 5. Schematic representation of the three types of galectins (proto-, chimera- and tandem-repeat). The proto-type is composed of only a single lectin domain. The chimera type is composed of two parts: a C-termainal half containing the galectin CRD and an N-terminal half of unknown function. The tandem-repeat type is composed of two homologous CRD domains. Above the proto-type structure, conservative residues are shown in *single letters. Large capital letters* denote the most important residues for sugar binding. *Asterisks* denote residues mostly conserved in all galectins studied so far. Residue numbers are those of human galectin-1. (Hirabayashi and Kasai 1988)

multicellular animals in order to carry out certain fundamental tasks, and thus became widely distributed throughout the animal kingdom.

4.2 Biochemical Properties of the Galectin Family

4.2.1 *General Properties*

Galectins are water-soluble proteins. However, for the initial process of purification, it is necessary to add some hapten sugars such as lactose to the extraction buffer in order to dissociate them from insoluble substances in the starting tissues. This does not mean they are integrated into lipid membrane. They appear to bind to sugar chains of insoluble glycoconjugates such as glycoproteins, glycolipids, and proteoglycans.

Since the sugar-binding activity of galectins was often lost in the absence of a reducing agent such as β mercaptoethanol, the presence of (a) free SH group(s) was once thought to be essential (Levi and Teichberg 1981). However, various experiments including chemical modification (Whitney et al. 1986) and site-directed mutagenesis (Hirabayashi and Kasai 1991) later showed that SH groups were not essential for sugar binding. This was also supported by the absence of cysteine residues in *C. elegans* 32-kDa galectin (Hirabayashi et al.

1992). Oxidation of SH groups caused a global conformational change which resulted in the loss of sugar-binding ability (Tracey et al. 1992). A series of recombinant galectins, in which each conserved amino acid residue located in the CRD region was substituted by site-directed mutagenesis, were also produced (Hirabayashi and Kasai 1991, 1994; Abbott et al. 1991). The residues proven to be indispensable for the sugar-binding activity are shown by block capital letters in Fig. 5. X-ray crystallography of galectins revealed that these residues were found to form hydrogen bonds with principal hydroxyl groups in lactose; 4-OH of Gal with His45, Asn47 and Arg49, 6-OH of Gal with Asn58 and Glu68, and 3-OH of Glc with Arg49, Glu68 and Arg70 (residue numbers are those for human galectin-2) (Fig. 6; Lobsanov et al. 1993; Bourne et al. 1994; Liao et al. 1994).

All galectins studied so far have blocked N-termini, and an acetyl group was shown to be the blocking group (Hirabayashi et al. 1987; Hirabayashi and Kasai 1988). Although galectins occasionally contain cysteine residues, no disulfide bond is formed, and all SH groups are in a free state. No galectins, except those of sponges, are glycosylated. Cloning of cDNAs revealed that galectins are synthesized without a signal sequence (Ohyama et al. 1986). (In the cDNA for sponge galectins, some hydrophobic sequences were found, but it has not been confirmed that it corresponds to the signal sequence or membrane integration signal; Pfeifer et al. 1993.) These characteristics strongly suggest that galectins are designed as intracellular proteins. However, galectins were very often found outside cells. Therefore, they must be externalized by an unknown mechanism that does not depend on the signal peptide (Cooper and Barondes 1990).

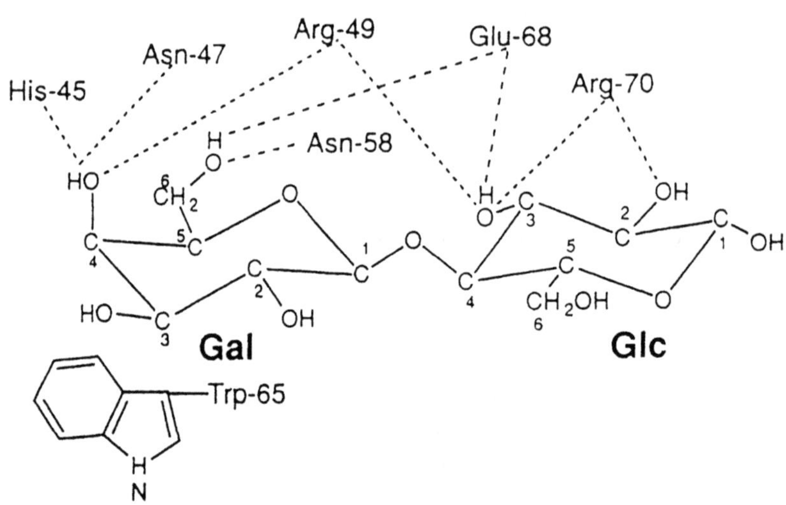

Fig. 6. Interaction between amino-acid residues in the galectin CRD and the lactose molecule revealed by X-ray crystallography. (Lobsanov et al. 1993)

4.2.2 *Three Types of Galectins*

As mentioned earlier, mammalian galectins are distinguished by serial numbers, but these numbers give us no significant information on their molecular properties and functions. Instead, it would be more practical to classify them according to their molecular architecture. Therefore, we proposed to classify them into proto- chimera, and tandem-repeat types (Fig. 5; Hirabayashi et al. 1992). Such a classification greatly helps us to investigate galectins of different origins.

The proto-type galectin comprises small proteins of about 15000 daltons containing only one CRD domain, and the most common one is mammalian galectin-1. In addition, mammalian galectins-2 (Gitt et al. 1992), -5 (Gitt et al. 1995), and -7 (Madsen et al. 1995; Magnaldo et al. 1995) have also been found to be proto-types. Of them, galectins-1, -2 and -7 exist as noncovalent homodimers under physiological conditions, whereas galectin-5 was reported to exist as a monomer. In chickens, two proto-type galectins (C-14 and C-16, named according to the apparent molecular weights deduced from SDS-PAGE) have been found (Hirabayashi et al. 1987; Sakakura et al. 1990). The chimera type, which has been found only in mammals (Jia and Wang 1988) and chickens (Nurminskaya and Linsenmayer 1996), has a molecular weight of about twice that of the proto-type (30-35kDa), and is composed of two different domains. The C-terminal half is the galectin domain containing one CRD. The N-terminal half, exhibiting no homology to galectins, is related to components of the heteronuclear ribonucleoprotein complex (hnRNP). Characteristic repetitive sequences rich in proline and glycine are also found in this type. *C. elegans* 32-kDa galectin, and mammalian galectins-4 (Oda et al. 1993), -6 (Gitt et al. 1995) and -8 (Hadari et al. 1995) belong to the tandem-repeat type. They contain two CRD domains in a single polypeptide chain.

From a functional viewpoint, these three types must behave differently. Proto-type galectins form divalent homodimers, which consequently will cross-link two glycoconjugates of a very similar nature. Tandem-repeat types are divalent, but the specificity and binding strength of the two binding sites are not necessarily the same (Oda et al. 1993; Arata et al. 1997). Therefore, they can form bridges between glycoconjugates of different types. Chimera types have one sugar-binding site and other unknown regions, which might interact with biomolecules other than sugars. Therefore, they can serve as cross-links between glycoconjugates and other biomolecules, in other words, as adaptor molecules between different kinds of biomolecules or as heterobifunctional cross-links.

4.3 Gene Structures of Galectins

Genomic structures have been determined for the first time for chicken 14kDa galectin, a proto-type galectin (Ohyama and Kasai 1988). Up to now, genes for

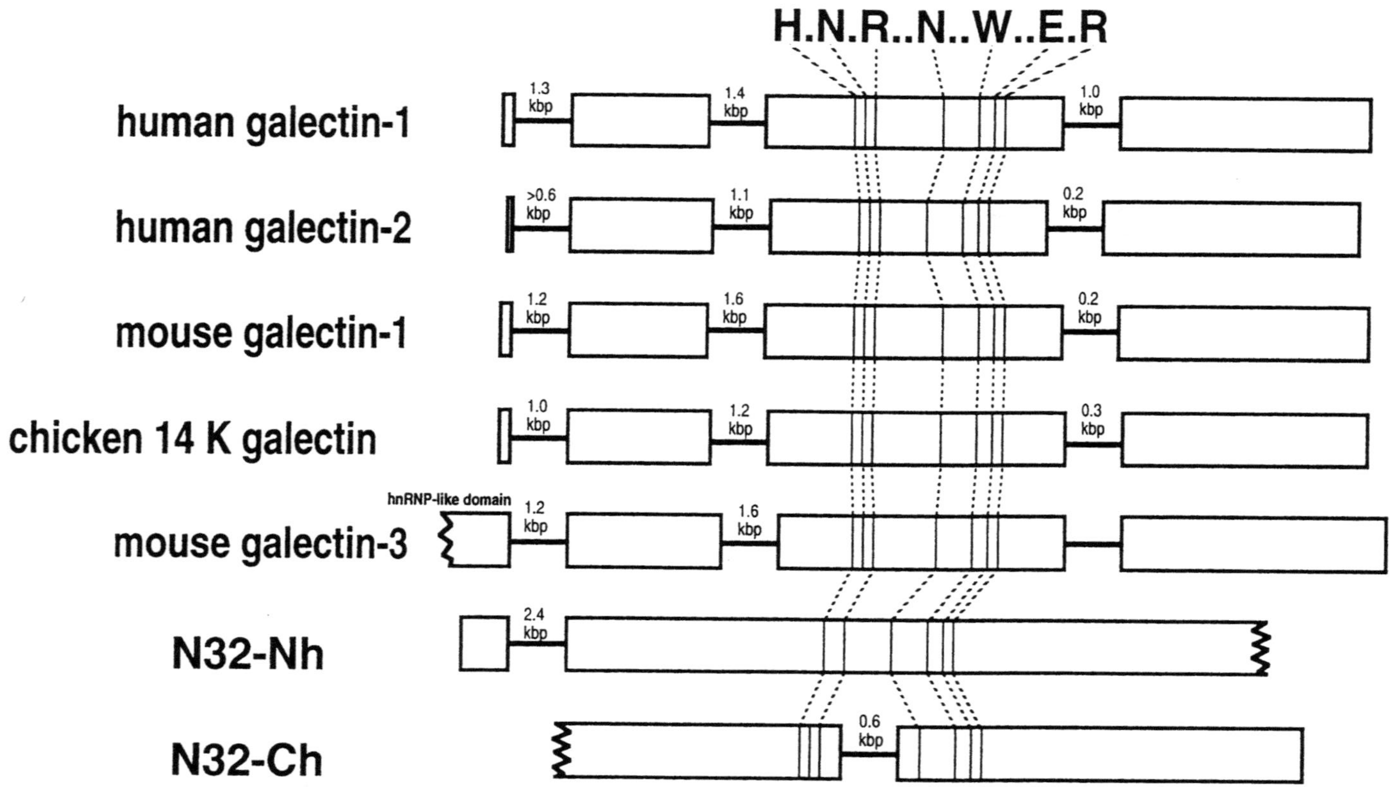

Fig. 7. Genomic structures of galectin genes. *Open boxes* stand for exons and lines for introns. The conserved amino acids (indicated *at the top* in *one-letter codes*) among the galectin CRDs are shown in *vertical lines connected by dotted lines*. The genomic structure of the 32-kDa galectin for the two tandemly repeated lectin domains (*Nh* and *Ch* stand for N-terminal half region and C-terminal half region, respectively) are shown *at the bottom*

human galectin-1 (Gitt and Barondes 1991) and -2 (Gitt et al. 1992), mouse galectin-1 (Chiariotti et al. 1991), mouse galectin-3 (Gritzmacher et al. 1992) and nematode 32-kDa galectin (Arata et al. 1997) have been cloned and analyzed (Fig. 7). Basic organization of the gene seems to be conserved in the case of vertebrate galectins, including not only the proto-types but also the chimera type; three introns are inserted at similar positions in the coding region. The coding region of vertebrate galectins consists of four exons: from the 5′ end to the 3′ end, they are exon I (6–9 bp), II (80–83 bp), III (160–172 bp), and IV (144–150 bp). The largest exon, III, encodes the most conservative, hydrophilic region including the amino acids essential in forming hydrogen bonds with the bound lactose molecule (amino acids indicated at the top of Fig. 6).

The exon-intron arrangement of the gene for the 32-kDa galectin of *C. elegans* is different (Arata et al. 1997). It contains two relatively long introns, namely, intron 1 and intron 2. Intron 1 is found near the 5′ end of the coding region, and intron 2, in the central region of the C-terminal half of the lectin domain. The insertion point for intron 1 is similar to that of vertebrate galectin genes, although this intron is extremely long in comparison with the corresponding vertebrate introns. Intron 2 is peculiar because it is inserted in the center of the CRD of the C-terminal half-region. Such a type of insertion is not observed in any other known galectin genes. Since the two repeated lectin domains of the nematode 32-kDa galectin would seem to have been created by gene duplication, the presence of an intron between the two lectin domains had been expected, but this is not the case.

4.4 Molecular Evolution of Galectins

4.4.1 *Sequence Similarity Between Galectins*

Galectins have been found in vertebrates (mammals, birds, frogs, and fish) and in invertebrates such as nematodes and sponges. Table 4 shows the degree of substitution of amino acids among some typical galectins. Galectin-1 is the most common galectin in mammalian species. The primary structures are very similar in galectin-1 isolated from various mammalian species, and about 90% of the amino acid residues are identical. As to galectins of nonmammalian species, it is not easy to correlate any one of them to any one of the mammalian galectins. For example, though two proto-type galectins (C-14 and C-16) have been found in chickens, neither is similar to any one of the mammalian galectins (Sakakura et al. 1990). Both chicken galectins differ from mammalian galectin-1 to approximately the same extent, i.e., the degree of amino acid substitution between C-16 and mammalian galectin-1 is 59–63 per 135 and that between C-14 and mammalian galectin-1 is 58–60 per 135. On the other hand, the degree of substitution between the two chicken galectins is 71

Table 4. Homology matrix for galectins lectins showing amino acid sequence identities (%) and *Kaa*[a]

	H14	C14	C16	H14-II	Xen14	Con14	N32-i	N32-ii	H30	S14-I	CLC
H14	-	57.9	57.1	43.4	39.7	33.1	27.3	25.4	26.5	18.5	20.3
C14	0.55	-	47.0	39.5	35.6	34.1	28.1	27.7	29.5	19.1	15.6
C16	0.56	0.75	-	39.5	32.6	34.8	27.3	26.9	24.2	16.0	22.7
H 14-Il	0.83	0.93	0.93	-	33.3	32.6	27.9	23.8	29.2	20.2	16.0
Xen14	0.92	1.03	1.12	1.10	-	25.8	22.0	24.2	22.1	21.5	13.5
Con14	1.11	1.08	1.06	1.12	1.10	-	28.1	21.7	22.9	17.7	15.6
N 32-i	1.30	1.11	1.30	1.28	1.51	1.27	-	33.3	30.8	17.1	16.2
N 32-ii	1.37	1.30	1.31	1.43	1.42	1.53	1.10	-	34.6	18.6	27.0
H30	1.33	1.22	1.42	1.23	1.51	1.47	1.18	1.06	-	19.8	25.6
S14-I	1.69	1.66	1.83	1.60	1.54	1.73	1.77	1.68	1.62	-	11.9
CLC	1.59	1.86	1.48	1.83	2.00	1.86	1.82	1.31	1.36	2.13	-

[a] Data above the diagonal represent amino acid sequence identities (%) between each pair of sequences using the alignment presented in Fig. 2. Gaps were not counted for calculation. Data below the diagonal represent amino acid substitutions per site, including possible multiple substitutions (*K*aa), using the equation *K*aa = −ln [observed identity (%)/100; Kimura 1983]. For lectin abbreviations, see legend to Fig. 2

per 134. This suggests that divergence of the two chicken proto-type galectins occurred a very long time ago, presumably close to the time of divergence of birds and mammals. Divergence of galectins seems to have occurred independently in birds and mammals. This is one of the reasons why the serial numbers given to mammalian galectins are not applicable to nonmammalian galectins.

Between mammalian galectins, galectin-2 has 43% sequence identity to human galectin-1 and is more similar to galectin-1 than to any other known galectins. The sequence of galectin-5 is about 40% identical to galectin-3 and -4 and about 20% identical to galectin-1 and galectin-2. It is also very similar to the second CRD of galectin-9 (70–85% identity). Similarity between two galectin domains of galectin-4 is about 40%, and they are similar to galectin-3 and galectin-5, but only about 20% identical to galectin-1 and -2. Two galectin domains of galectin-6 are about 80% identical to those of galectin-4. Therefore, there seem to be two groups in mammalian galectins. One is composed of galectin-1 and -2, and the other, galectin-3, -4, -5, -6, -9 (Gitt et al. 1997). On the other hand, Ahmed and Vasta (1994) proposed classifying galectins based mainly on sugar-binding specificity. According to them, galectin-1 belongs to type I (conserved) and galectin-2, -3, and 4 form type II (variable). More detailed analysis from various points of view will be needed to elucidate the evolutional relationship between mammalian galectins.

4.4.2
Divergence of Galectins

If we assume that the time of divergence of mammalian species was 300×10^6 years ago, the rates of amino-acid substitution for each pair of galectins can be calculated. The obtained values were in the range of $0.6–1.1 \times 10^{29}$ substitutions per site per year (Sakakura et al. 1990). These values are close to the substitution rate for the α-chain of hemoglobin, suggesting that galectin-1 has been as conservative as hemoglobin and thus plays some role as important as hemoglobin. If we assume that these values are applicable to all vertebrate galectins, the time of divergence of the two chicken galectins (C-14 and C-16) can be calculated. The result suggested that C-14 and C-16 were separated 380 $\times 10^6$ years ago. The differentiation of these two chicken galectins seems to have occurred at a time not so distant from the divergence of birds and mammals (Fig. 8).

Galectins are now known to form a much larger family than had been previously thought. It seems likely that virtually all animal species have some proteins belonging to this family, because an ancestral protein of vertebrate galectins and nematode galectins would have existed before the divergence of the two big branches of the phylogenic tree, i.e., the protostomes and deuterostomes. If we assume that the same amino-acid substitution rate calculated for vertebrate galectin-1 can be applied, the divergence time between nematode and vertebrate galectins can be estimated to be 6.8×10^8 years ago,

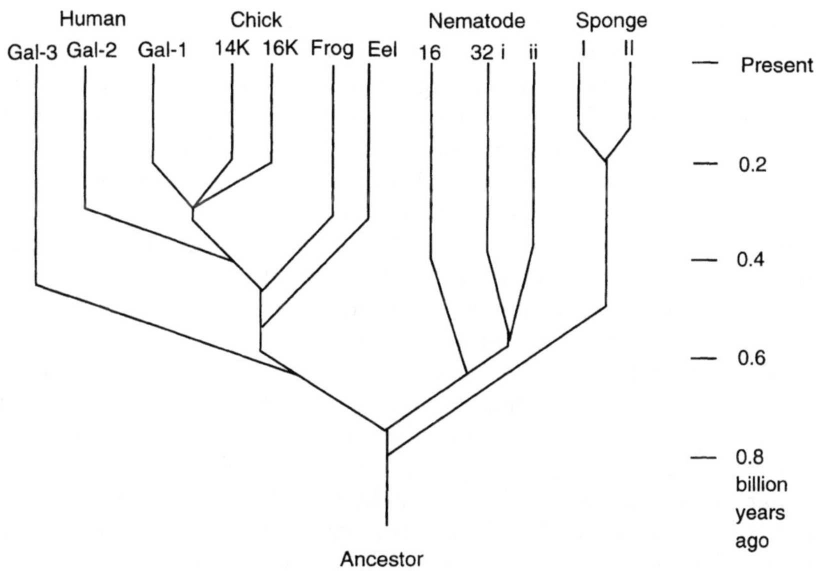

Fig. 8. A phylogenic tree for typical metazoan galectins

that is, in the Precambrian era (Hirabayashi et al. 1992). This estimation agrees well with the generally accepted time of the divergence of deuterostomes and protostomes. On the other hand, divergence of the two repeated domains of the nematode 32-kDa galectin was estimated to have occurred 5.6×10^8 years ago. This suggests that the formation of the repeated structure by gene duplication occurred after the divergence between nematodes and vertebrates. This explains why no apparent correlation is found between the nematode 32-kDa galectin and mammalian tandem-repeat galectins. Gene duplication and fusion of the proto-type galectin, which resulted in the birth of tandem-repeat galectins, seem to have occurred independently in both nematodes and mammals.

Since chimera-type galectins have been found only in vertebrates and none have been found in *C. elegans*, it is not possible to guess at which stage of animal evolution they were created. Were they created by combining the galectin and another functional domain of a certain protein after vertebrates appeared on the earth? It was reported that the N-terminal half of the chimera-type galectin (i.e., galectin-3) is homologous to some nuclear proteins and that galectin-3 is one of the components of the splisosome and plays an important role in splicing pre-mRNA (Anderson and Wang 1992; Dagher et al. 1995). If galectin-3 is involved in such a fundamental process as maturation of mRNA, it should be found in a much wider range of animal species.

The fact that both nematodes and vertebrates have galectins indicates the existence of a common ancestor before the divergence of protostomes and deuterostomes. Since sponges are more primitive, there should also have been a common ancestor between their galectins and other metazoan galectins. The time of divergence was calculated to be 800×10^6 million years ago, by assuming the same amino-acid substitution rate. On the other hand, two sponge galectins found so far seem to have diverged only 220×10^6 years ago, because they are about 65% similar (Pfeifer et al. 1993). The discovery of sponge galectins has pushed back the origin of the galectin family nearly to the start of multicellular animals. The fundamental roles of the galectin family do not seem to have changed significantly from that time, because their basic structures have been conserved. These findings also lead us to expect to find galectins in all metazoan animals, although, at present, galectin has been found only in vertebrates, nematodes, and sponges.

Since the highest deuterostomes, i.e., vertebrates, still require galectins, lower deuterostomes such as the chordata and echinodermata would be expected to preserve them also. Even if they have ceased to use galectins for some reason, it is very possible that genetic information for galectins still remains as either silent genes or pseudogenes in them. It also seems rather strange that no galectin has been found in the arthropoda, the biggest and most well-adapted group of protostomes. If they have abandoned galectins, many questions arise, such as why? when? and what are the replacements?

4.5 Galectins: Relatives of Legume Lectins?

Recent crystallographic analysis revealed that galectins have the so-called jelly-roll topology, i.e., a sandwich composed of two β-sheets (Fig. 9; Lobsanov et al. 1993). This is very similar to that of legume lectins such as concanavalin A (ConA) in spite of the absence of similarities in the primary structures. The spatial arrangement of subunits and also the relative positions of the two sugar-binding sites in the dimeric molecule are also close. Some crystallographers have supposed that the number of topologies of existing proteins is not infinite but less than 1000. If this is the case, the topological resemblance between the sugar-binding proteins of legumes and animals might be the result of convergent molecular evolution, though this is not fully persuasive. It seems more plausible that an ancestor protein related to sugar molecules somehow, having a jelly-roll topology, existed on the primitive earth before the divergence of plants and animals. Even after the divergence, this topology was conserved, and molecular evolution gave rise to legume lectins in plants and galectins in animals, although the accumulation of point mutations has made it impossible to find sequence similarities today.

The search for proteins linking legume lectins and galectins will need a new strategy, because the homology in primary structure is too low, and the only

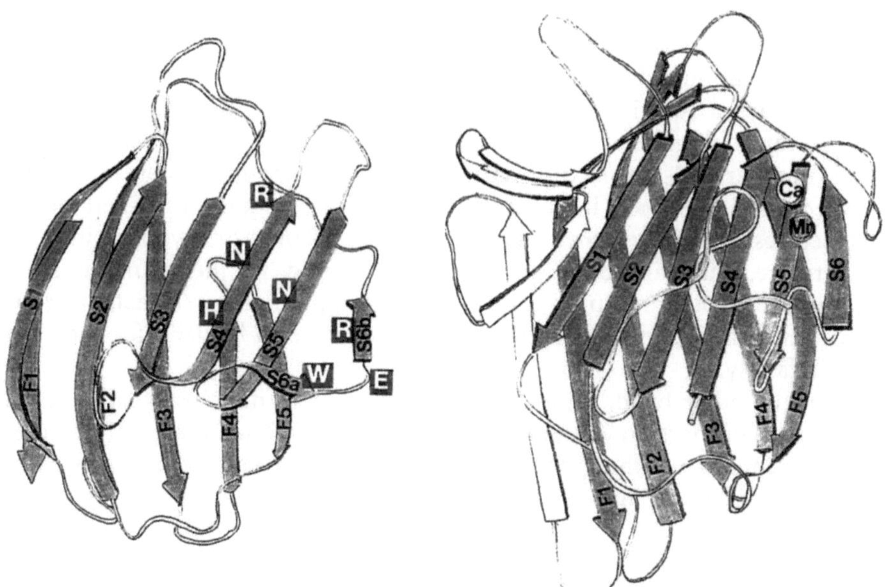

Fig. 9. Comparison of the topologies of human galectin-2 (*left*) and concanavalin-A (*right*), modified from a figure drawn by Richardson (1981). β-strands in concanavalin A corresponding to those in galectin-2 are *shaded*. Conserved amino acid residues (capital letters) are shown in galectin-2, and the positions of metal ions (*Ca and Mn*) are indicated in concanavalin A

clue is their topology. However, recent observations indicate that the galectin family is now growing into a galectin super family: the Charcot-Leyden crystal protein, which exhibits lysophospholipase activity, was found to exhibit sequence homology to galectins (Ackerman et al. 1993); crystallography revealed that pentraxin, a serum amyloid P-component, has a topology similar to that of galectins (Emsley et al. 1994); and very recently, the presence of galectins in some species of fungi has also been reported (Cooper et al. 1997).

5 Comparison of the Galectin Family and the C-Type Lectin Family

It is interesting that there are such sharp contrasts between the galectin family and the C-type lectin family, although both families are principal animal lectins. It seems very helpful for a better understanding of the biological meaning of sugar recognition to consider the possible reasons why they are so different.

5.1 Family Size

The C-type lectin family is very large in comparison with the galectin family. The number of C-type lectins in mammals has already exceeded 50, and new members are being continually found. Each C-type lectin is usually specialized for one or a limited number of well-defined functions. For example, hepatic lectin plays a role in removing old serum proteins, thrombomodulin is a regulator of blood clotting, selectins guide leucocytes, and so on. A considerable number of C-type lectins were regarded as functional proteins rather than as lectins. Sometimes, it was only after the determination of their primary structures that they turned out to be members of the C-type lectin family. Therefore, they are often given names related to their functions, e.g., asialoglycoprotein receptor, thrombomodulin, pulmonary surfactant protein, etc. The "one lectin-one function" concept seems to be valid for the C-type lectin family.

On the other hand, a much smaller number of members of the galectin family are in charge of extremely diverse processes. Therefore, each galectin is not sepcialized and thus has to carry out multiple tasks. It is rather rare that a galectin is given a name related to its function.

5.2 Molecular Architecture

The molecular architectures of C-type lectins reflect the diversity of their functions. They are usually composed of the characteristic C-type lectin domain (CRD), combined with certain functional domains of other proteins (such as growth-factor motif, membrane-binding domain, collagen domain, etc.). The C-type lectin family seems to be open and advanced, and the creation of new members has been allowed, as the life systems became more and more complicated and a number of new special molecules were required. New members were probably created frequently by combining the CRD with an appropriate part of other functional proteins.

In contrast, the variety in molecular architecture in the galectin family is poor; and only three types have been found, i.e., proto, chimera, and tadem-repeat types. The galectin family seems to have been closed and conservative, with the family size kept at a minimum and multiple tasks handled by a limited number of family members. The functions now carried out by the galectin family seem to have origiated in a very early period of life on earth and have not changed essentially during evolution. They seem to be more general and fundamental than those carried out by C-type lectins.

5.3 Binding Specificity

The sugar-binding specificity of the C-type lectin family is wide, and various sugar moieties such as GlcNAc, Man, Gal, and Fuc, etc., are recognized by different C-type lectins. Alterations of the sugar-binding specificity were permitted during their molecular evolution. On the other hand, galectins have been conservative in their sugar-binding specificity also. The sugar-binding specificity of galectins has been analyzed by different groups (Leffler et al. 1986; Abbott et al. 1988; Lee et al. 1990; Ahmed and Vasta 1994; Gupta and Brewer 1994). No galectin specific for saccharides other than galactosides has been found. Why is the sugar-binding specificity of the galectin family so monotonous? It is possible that the change in the specificity has been restricted because the recognition of certain glysocodes by galectins has been kept as a basic process even though Metazoa have evolved into extremely complicated systems.

5.4 Localization

The structural characteristics of the galectin family (i.e., acetylated N-terminus, and the lack of disulfide bonds, signal peptide sequences, and saccharide chains) clearly indicate that they are designed as intracellular proteins. On the other hand, C-type lectins are apparently designed to function extracellularly. This suggests fundamental differences in the origin, history, and roles of these two families. Such fossil evidence that exists suggests that the ancestor of galectins was initially created for roles played inside primitive unicellular organisms, and their offspring acquired new roles played outside cells after multicellular organisms appeared and communication between cells became more and more important. On the other hand, C-type lectins seem to have been utilized from the beginning for the purpose of communication carried out outside cells in multicellular organisms.

5.5 Different Missions Assigned to Galectins and C-Type Lectins

The above differences lead to the assumption that the missions assigned to galectins are very old and general, whereas those given to C-type lectins are newer and specific. This is a possible explanation as to why galectins are modest, conservative, and less highlighted, having limited binding specificity and a small family size. However, they are flexible and able to participate in multiple processes. On the other hand, C-type lectins have been more radical: the family has expanded continually in response to the increasing number of new biological processes. They became more and more specialized and at the same time very bureaucratic. As a result, each can now carry out only a limited process.

To operate the living systems appropriately, in addition to a great number of bureaucratic specialists, a small number of flexible generalists is also indispensable. Therefore, the galectin family could not have changed their basic characteristics. Galectins have thus kept their nature as nonbureaucratic bureaucrats, intelligent glue, or almighty supporting actors throughout the evolution of Metazoa.

6 Evolution of Glycocodes

6.1 Galactose: A Special Position in Sugar Recognition

Lectins are proteins that recognize the information presented by sugar chains. In other words, they decipher glycocodes. In this section, we will discuss the origin and evolution of the elements that construct the glycocodes. For this purpose, we first examine which sugars are used most frequently as the counterparts of lectins.

Goldstein and Poretz (1986) classified 56 lectins, most of which are of plant origin. They were classified into five groups based on monosaccharide specificity, i.e., groups of lectins specific for Gal/GalNAc (30 cases, 54%), α-Glc/Man (10 cases, 18%), GlcNAc (9 cases, 16%), L-Fuc (4 cases, 7%), and sialic acid (3 cases, 5% ; Table 5). More recently, Doyle (1994) have collected as many as 237 lectins, including a higher proportion of animal lectins with defined specificity, and the above result was confirmed: 145 Gal/GalNAc-specific lectins (61%), 33 Glc/Man-specific lectins (14%), 28 GlcNAc-specific lectins (12%), 16 L-Fuc-specific lectins (7%), and 15 sialic acid-specific lectins (6%) (Table 5). Obviously, lectins specific for galactosides are predominant.

Table 5. Classification of lectins on the basis of simple saccharide specificity

I. by Goldstein and Poretz (1986)

	Gal/GalNAc	Glc/Man	GlcNAc	L-Fuc	Sialic acid	Total
Animal	1	0	0	1	3	5
Plant	29	10	9	2	0	50
Microorganism	0	0	0	1	0	1
Total	30(54%)	10(18%)	9(16%)	4(7%)	3(5%)	56

II. by Doyle (1994)

	Gal/GalNAc	Glc/Man	GlcNAc	L-Fuc	Sialic acid	Total
Animal	41	1	2	3	14	61
Plant	93	28	24	8	1	154
Microorganism	11	4	2	5	0	22
Total	145(61%)	33(14%)	28(12%)	16(7%)	15(6%)	237

Many animal lectins, especially galectins and C-type lectins, which were included in Doyle's list, show apparent galactose recognition. Although the specificities of C-type lectins are rather diverse, all galectins found so far are specific for galactosides. This means that the specificity of galectins has been conserved during the extremely long period of molecular evolution. These observations suggest that galactose occupies some special position in the sugar-recognition phenomena. Why is galactose preferentially utilized as a critical epitope recognized by lectins? And what distinguishes galactose from other elementary sugars such as glucose and mannose? We will now try to find a possible explanation based on physicochemical and biochemical data.

The key structure recognized by most galectins seems to be disaccharide unit, lactose (Gal β1-4Glc) and N-acetyllactosamine (Gal β1-4GlcNAc, LacNAc) (Leffler and Barondes 1986; Abbott et al. 1988; Mandal and Brewer 1992; Satoh and Hughes 1992; Gupta and Brewer 1994). These structures are included in a variety of glycoconjugates and also serve as a backbone for various glycosignals such as blood-group determinants and Lewis antigens. Polymerized LacNAc structures, polylactosamine chains, are also often found in glycoconjugates, such as fibronectin, laminin, etc., and have been suggested as candidates for endogenous receptors for galectins (Oda and Kasai 1984; Do et al. 1990; Zhou and Cummings 1993; Ozeki et al. 1995). Therefore, Lac and LacNAc units seem to be one of the most important glycocodes and to have originated from the time when the ancestor of galectins appeared on the earth. Although the affinity of galectins for various galactoside-containing oligosaccharides differs slightly from one galectin to another, the fundamental specificity for Lac and LacNAc seems to have been preserved. The deciphering or decoding of these glycocodes by the galectin family has been continually one of the most important processes in living organisms (Kasai and Hirabayashi 1996).

Galactose recognition seems to be more common in multicellular organisms than in primitive microorganisms. Prokaryotes usually seem to prefer recognizing systems using mannose or glucose. This can be related to the fact that multicellular organisms have developed complex-type glycoconjugates containing galactose, whereas most prokaryotes usually have glycoconjugates composed of mainly mannose and glucose. This suggests that the recognition of glycocodes containing galactose moieties became more common after multicellular organisms appeared. Before that time, sugar recognition might have been based mainly on glucose and mannose.

In higher eukaryotes, galactose is located at nonreducing termini of glycoconjugate sugar chains, often modified with sialic acids and L-fucose. On the other hand, a so-called "core structure"; i.e., Manα1-3(Manα1-6)Manβ1-4GlcNAc1-4GlcNAc-Asn of N-glycans, is conserved between phylogenically distant species, i.e., from yeasts to mammals. This implies that *N*-glycan biosynthesis was developed after the divergence of archeobacteria and eukaryotes.

Galactose is distinct from glucose/N-acetylglucosamine and mannose in that its 4-OH group is axial in its preferred chair conformation (Fig. 10). Not only the presence of the 4-axial hydroxyl group but also the nonreducing terminal location gave galactose a unique status among elementary saccharides, i.e., as a recognition molecule. The 4-hydroxyl group has been proved to be most important for the recognition of galactose by various galactose-specific lectins. The 2-hydroxyl group of mannose, which has axial configuration in the assumed C1 conformation, is not required for recognition by most mannose-specific lectins (Derewenda et al. 1989; Bourne et al. 1990; Weis et al. 1992). This explains why most mannose-specific legume lectins and animal lectins (described in Sect. 3.6.2) can also bind glucose. Therefore, we will consider why galactose has acquired such a position in constructing the glycocodes.

6.2 Selection of Elementary Sugars in the Life System

First, we will consider why only three of the eight possible aldohexoses of the D-configuration (glucose, mannose, and galactose) are present in nature. The selection of the three elementary sugars seems to have been based on their thermodynamic stability. Glucose is the most stable aldohexose and plays a central role in both body construction and energy metabolism.

The thermodynamic stability of cyclohexane-type compounds, which are assumed to take more favorable chair forms rather than boat forms, is often discussed based on the concept of 1,3-diaxial interaction. For example, the C1-chair conformation of β-glucose has all five substituents (i.e., 1, 2, 3, 4-hydroxyl groups and the 5-hydroxymethyl group) in equatorial configura-

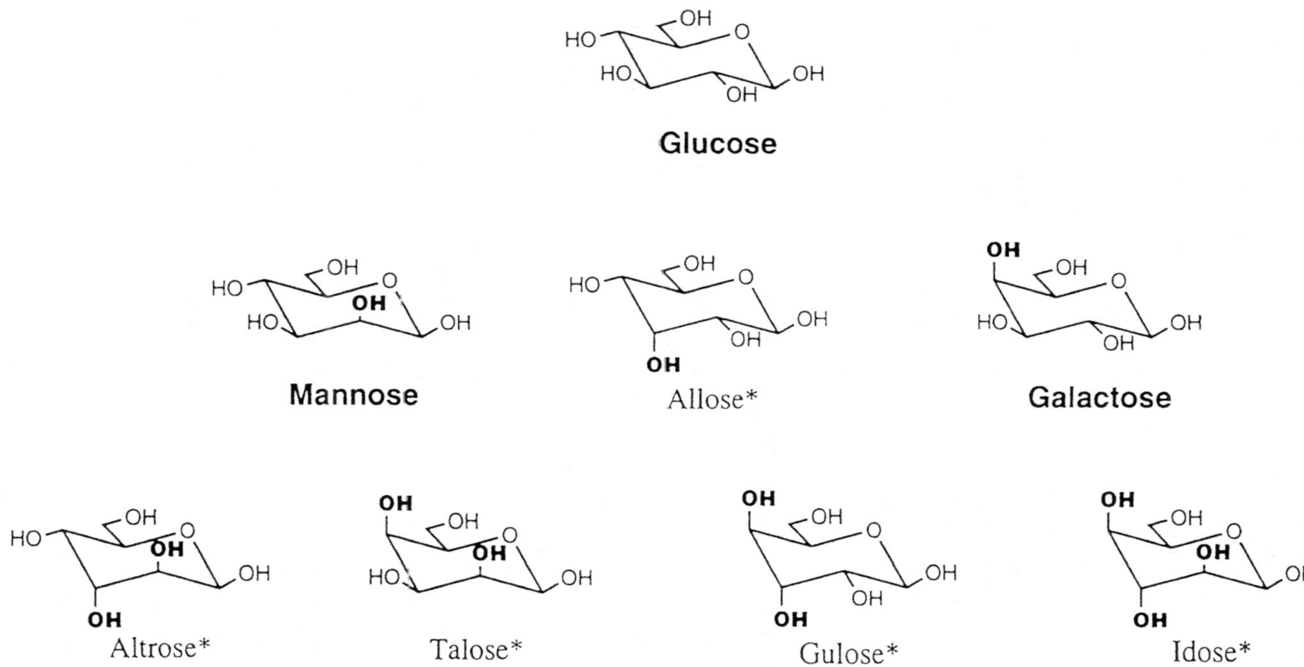

Fig. 10. Structures of the eight possible aldohexoses. For systematic comparison, all of them are drawn in β-configurations of C1-chair forms. Axial hydroxyl groups are drawn in *boldface* to emphasize the differences from β-glucose. *Asterisks* Allose, altrose, talose, gulose, and idose are not found in nature, except that D-talose has been found as a component of an antibiotic, hygromycin B

tions. It has no 1,3-diaxial interactions and thus is most stable from a thermodynamic viewpoint. On the other hand, both mannose and galactose have a single axial hydroxyl group at the 2- and 4-positions, respectively (Fig. 10). Therefore, they can be regarded as 2-epi-glucose and 4-epi-glucose, respectively. As a result, however, they have one (but not two) 1,3-diaxial interaction, i.e., between 2-OH (axial) and 4-H in mannose, and between 2-H and 4-OH (axial) in galactose. Due to such thermostability, these three monosaccharides seem to have been selected as elementary molecules for life. The remaining glucose epimer, allose (i.e., 3-epi-glucose), has two such interactions, i.e., between the 1–3 and 3–5 positions, and thus is less stable. In other words, allose cannot escape an extra interaction in the way that mannose and galactose can (the latter two molecules have an oxygen atom at one of the two 1–3 positions). This may explain why allose was not selected in nature.

None of the glucose diastereomers, i.e., altrose, gulose, idose, and talose, are found in nature, probably because they are much less stable.

6.3
Monosaccharides: Biosynthesis from Glucose and Mannose

From the biosynthetic viewpoint, almost all saccharides are derived from glucose and mannose (Fig. 11). In contrast, there are no cases where galactose is utilized as a starting material. It is well known that glucose, mannose, and fructose are interconvertible (Fig. 12). This type of conversion proceeds via a linearized enediol 6-phosphate intermediate in the presence of a specific isomerase, but without coenzyme. Glucose 6-phosphate is converted to fructose 6-phosphate by the action of glucose-phosphate isomerase before entering the glycolytic pathway. Notably, the reaction mechanism apparently mimics that of the Lobry de Bruyn rearrangement (Speck 1958), a classic glycochemical reaction that can occur in the absence of enzymes under weakly basic conditions. The reaction utilizes basic chemistry, i.e., keto-enol tautomerism involving the 1-aldehyde group. Therefore, from a biosynthetic viewpoint, glucose and mannose play a central role in producing a variety of monosaccharides, including very unusual ones. On the other hand, galactose, though one of the elementary saccharides, does not seem to be utilized as a starting material, and may rather be regarded as a secondary product. Among the three elemental monosaccharides, galactose seems to be basically distinct from glucose and mannose.

6.4
A Possible Scenario of the Origin of Carbohydrates

Now we present a hypothesis that can explain the formation of the elementary monosaccharides under the conditions on the primitive earth and also their selection by primitive life (Hirabayashi 1996).

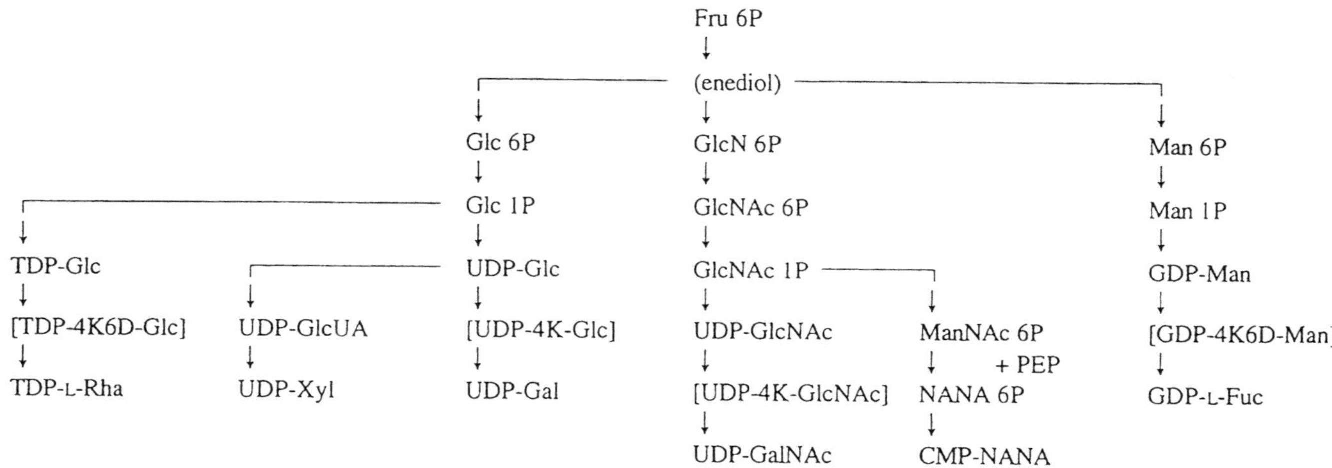

Fig. 11. Summary of the biosynthetic pathway for common saccharides. Data basically from Sharon (1975), but adapted to more clearly illustrate that fructose takes a fundamental position in saccahride biosynthesis. Abbreviations: *Fru 6P*, *Glc 6P*, *Man 6P*, *GlcN 6P*, *GlcNAc 6P*, *ManNAc 6P*, and *NANA 6P* are 6-phosphate derivatives of fructose, glucose, monnose, glucosamine; *N*-acetylglucosamine, *N*-acetylmannosamine, and *N*-acetylneuramic acid, respectively. *Glc 1P*, *Man 1P*, and *GlcNAc 1P* are 1-phosphate derivatives. *UDP-GlcUA* UDP-gluclonic acid; *PEP* phosphoenolpyruvate. *Bracketed compounds* are biosynthetic intermediates: *TDP-4K6D-Glc* TDP-4-keto-6-deoxyglucose; *UDP-4K-Glc* UDP-4-keto-glucose; *UDP-4K-GlcNAc* UDP-4-keto-*N*-acetylglucosamine; *GDP-4K6D-Man* GDP-4-keto-6-deoxy-mannose

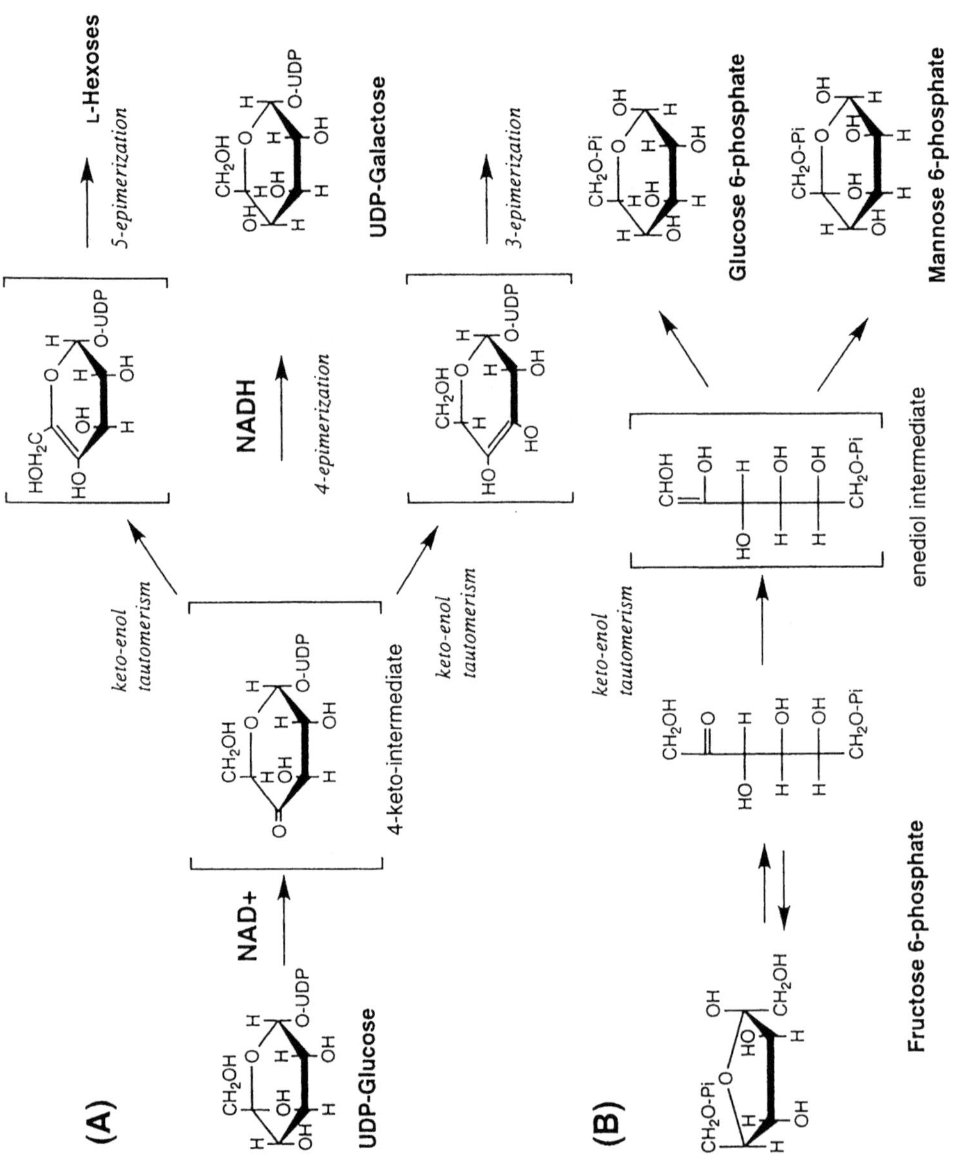
(A)
UDP-Glucose
NAD+
4-keto-intermediate
keto-enol tautomerism
keto-enol tautomerism
NADH
4-epimerization
UDP-Galactose
5-epimerization
L-Hexoses
3-epimerization
(B)
Fructose 6-phosphate
keto-enol tautomerism
enediol intermediate
Glucose 6-phosphate
Mannose 6-phosphate

Formol reaction has long been supposed to be a key reaction in understanding the origin of carbohydrates (Stanek et al. 1963; Decker et al. 1982; Miller 1987; Fig. 13). Formaldehyde (CH_2O) spontaneously polymerizes under relatively mild alkaline conditions into various small saccharides, including some important molecules such as glyceraldehyde and dihydroxyacetone. An important feature of the formol reaction is that it is usually accompanied by two conversion reactions, i.e., aldol condensation and Lobry de Bruyn rearrangement.

Once glyceraldehyde and dihydroxyacetone were produced, fructose would be produced very easily as the primary product of aldol condensation:

$$\text{glyceraldehyde (GA)} + \text{dihydroxyacetone (DHA)} \rightarrow \text{fructose.} \qquad (1)$$

Notably, this reaction is quite similar to the reverse of one of the reactions of glycolysis; i.e., cleavage of fructose 1, 6-bisphosphate [Eq. (2)]:

$$\begin{aligned}&\text{fructose-1,6-bisphosphate (F-1,6-P)} \rightarrow \\ &\quad \text{glyceraldehyde-3-phosphate (GA3-P)} + \\ &\quad \text{dihydroxyacetone-phosphate (DHA-P)} \quad \Delta G' = +5.5\,\text{kcal/mol.} \qquad (2)\end{aligned}$$

This reaction has positive ΔG value, meaning that the equilibrium favors the formation of hexose. If both components are in sufficient concentration, products will be formed readily. It is also important that the presence of water does not prevent formation of hexoses, because dehydration is not necessary for condensation. The above aldol condensation yields four possible ketohexoses differing in the configuration of the 3- and 4-hydroxyl groups, i.e., *ribo-*, *arabino-*, *xylo-*, and *lyxo-* hexuloses. In fact, however, 3,4-*trans* isoforms, i.e., *arabino-*(fructose-) and *xylo-*(sorbose-) hexuloses, became dominant, because such 3,4-*trans*-isoforms are more stable.

Once fructose was formed, two aldohexoses, glucose and mannose, are thought to have been generated by Lobry de Bruyn rearrangement (more formally, Lobry de Bruyn-Alberta van Ekenstein transformation; reviewed by Speck 1958) from fructose. The reaction easily proceeds under weakly alkaline conditions via an enediol intermediate (Fig. 12). On the other hand, such a rearrangement from sorbose gives less stable aldohexoses, i.e., gulose and idose. Therefore, the chemical synthesis not only of ketose (fructose) but also of aldoses (glucose and mannose) as a result of aldol condensation and subsequent Lobry de Bruyn rearrangement seems to have deep

Fig. 12A-B. Two types of epimerization reactions. **A** This type of epimerization occurs via the 4-keto-intermediate of nucleotide sugars with the aid of NAD^+ (or $NADP^+$) and is utilized for 3, 4 or 5-epimerization. For instance, UDP-galactose is derived from UDP-glucose by the action of UDP-galactose epimerase in the presence of NAD^+. **B** The other type of epimerization proceeds via a common linearized enediol 6-phosphate intermediate, from which fructose, glucose, and mannose are generated. The reaction is closely related to a classic glycochemical reaction, Lobry de Bruyn rearrangement, and does not require the presence of a coenzyme

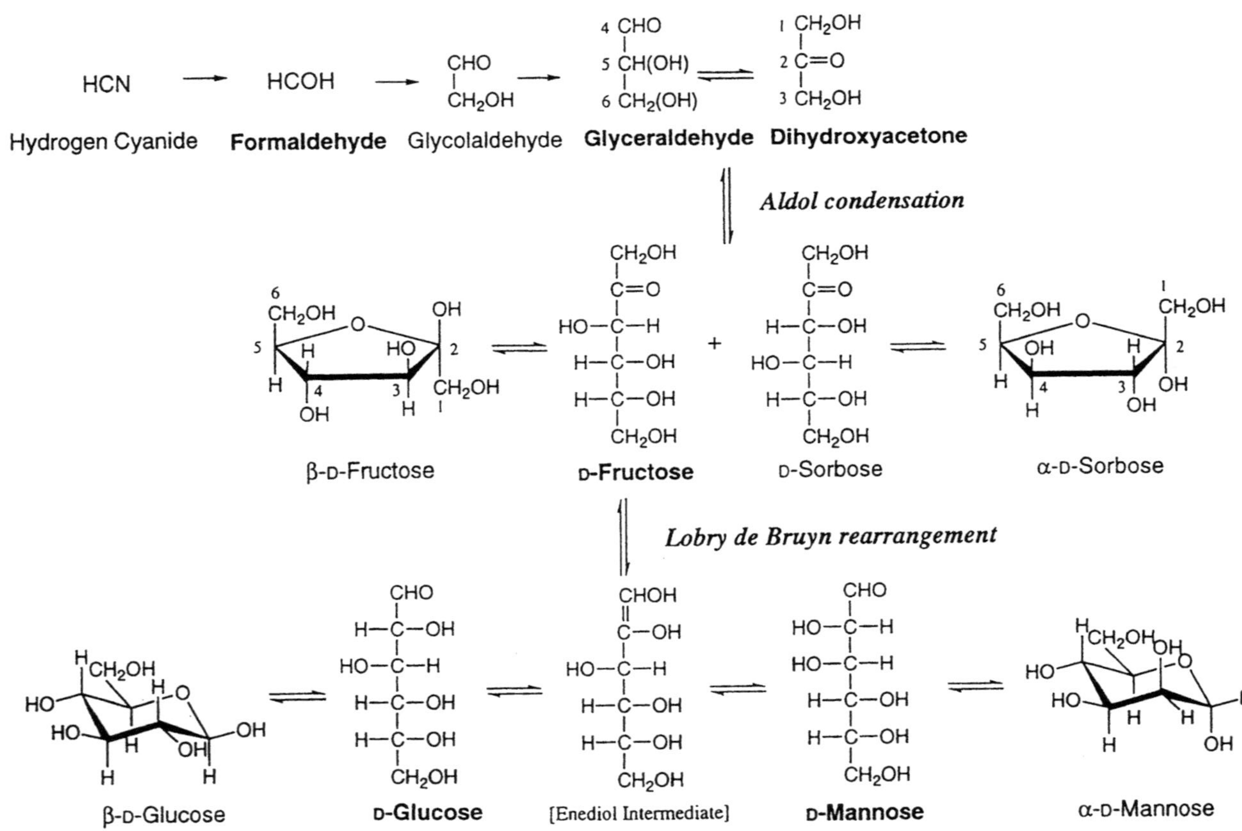

Fig. 13. Proposed scenario on the birth of elementary hexoses. The reactions to generate elementary hexoses consist of three consecutive steps: (1) formol condensation to generate glycelaldehyde and dihydroxyacetone from formaldehyde under weakly basic conditions; (2) aldol condensation between glycelaldehyde and dihydroxyacetone to generate 3,4-*trans*-ketoses (fructose and sorbose); and (3) Lobry de Bruyn rearrangement, which converts only fructose into glucose and mannose. These reactions can proceed in an equilibrium manner under weakly basic conditions. Note that galactose is not generated by the presented scenario, whereas fructose, glucose, and mannose form the first triplet hexoses. All of the hexoses presented here are shown in D-form for simplicity

meaning from the viewpoint of the creation and molecular evolution of saccharides.

It is also important to note that the above three sugars are in an equilibrium under weakly alkaline conditions. Therefore, if one of the components is removed, more will be supplied. In this hypothesis, however, galactose is not generated. Formation of galactose by Lobry de Bruyn rearrangement necessitates the preexistence of tagatose, from which talose should also be generated. However, no known organism utilizes tagatose, and galactose is produced from UDP-glucose via 4-keto formation with the aid of NAD^+. Such a galactose-synthesizing system was probably developed much later during the course of biological evolution. Apparently, the development of a biosynthetic system, i.e., the 4-oxide-reduction pathway and supply of a coenzyme, was necessary for the generation of galactose. Galactose seems to have become available when primitive living organisms appeared, and it became one of the recognition molecules at the time when multicellular eukarytes evolved, i.e., at the end of the Precambrian era. In other words, nature's selection of the first triplet, fructose-glucose-mannose, is based on a priori chemical reasons, but promotion of galactose to a recognition molecule occurred after life systems reached a certain level. There must have been some advantages for the usage of galactose as a component of glycocodes in multicellular systems. Therefore, galactose is a late-comer among the elementary monosaccharides.

Considering this glycochemical background, both glucose and mannose represent the most fundamental saccharides for microorganisms, i.e., bacteria, fungi, and protozoans. These organisms make maximal use of glucose and mannose, but not galactose, as starting materials for glycoconjugates. On the other hand, galactose has come to be utilized extensively in multicellular organisms for the purpose of recognition. Galactose has two inherent advantages over glucose and mannose as a recognition molecule, i.e., firstly, possession of a distinctive axial hydroxyl group (4-OH) at the most distant position from the core and, secondly, its outermost location in glycoconjugates as a result of arriving late on the scene. Consistent with this idea, the carbohydrate-binding proteins (lectins and carbohydrate-specific antibodies) studies so far have proved to be predominantly galactose-specific, and for them, without exception, the 4-hydroxyl group of galactose is essential for recognition. On the other hand, many mannose-specific lectins do not discriminate glucose from mannose but do distinguish it from galactose. Notably, animal galactose-specific lectins in general have endogenous ligands (e.g., poly-*N*-acetyllactosamine for galectins), whereas mannose-specific lectins (e.g., a group of collectins and mannose receptors) often recognize exogenous mannose-rich ligands expressed on pathogenic microorganisms. Therefore, from both theoretical and phenomenological viewpoints, galactose can be regarded as a late-comer saccharide distinct from other elementary hexoses. Possibly, utilization of galactose as a recognition molecule evolved concomitantly with the evolution of both complex multicellular organisms and lectins.

References

Abbott MW, Feizi T (1991) Soluble 14-kDa β-galactoside-specific bovine lectin: evidence from mutagenesis and proteolysis that almost the complete polypeptide chain is necessary for integrity of the carbohydrate recognition domain. J Biol Chem 266: 5552–5557

Abbott WM, Hounsell EF, Feizi T (1988) Further studies of oligosaccharide recognition by the soluble 13-kDa lectin of bovine heart muscle. Biochem J 252: 283–287

Abbott WM, Mellor A, Edwards, Feizi T (1989) Soluble bovine galactoside-binding lectin cDNA reveals the complete amino acid sequence and an antigenic relationship with the major encephalitogenic domain of myelin basic protein. Biochem J 259: 283–290

Ackerman SJ, Corrette SE, Rosenberg HF, Bennett JC, Mastrianni DM, Nicholson-Weller A, Weller P, Chin DT, Tenen DG (1993) Molecular cloning and characterization of human eosinophil Charcot-Leyden crystal protein (lysophospholipase). J Immunol 150: 456–468

Ahmed H, Vasta GR (1994) Galectins: conservation of functionally and structurally relevant amino acid residues define two types of carbohydrate recognition domains. Glycobioilogy 4: 545–549

Akimoto Y, Kawakami H, Oda Y, Obinata A, Endo H, Kasai K, Hirano H (1992) Changes in expression of the endogenous β-galactoside-binding lectin of chick embryonic skin during epidermal differentiation. Exp Cell Res 199: 3–18

Akimoto Y, Obinata A, Hirabayashi J, Sakakura Y, Endo H, Kasai K, Hirano H (1995) Changes in expression of two endogenous β-galactoside-binding isolectins in the dermis of chick embryonic skin during development in ovo and in vitro. Cell Tissue Res 279: 3–12

Albrandt K, Orida NK, Liu FT (1987) An IgE-binding protein with a distinctive repetitive sequence and homology with an IgG receptor. Proc Natl Acad Sci USA 84: 6859–6863

Anderson RL, Wang JL (1992) Carbohydrate binding protein 35. Trends Glycosci Glycotechnol 4: 43–52

Arata Y, Hirabayashi J, Akimoto Y, Hirano H, Kasai K (1996) An immunohistochemical study of the 32-kDa galectin (β-galactoside-binding lectin) in the nematode *Caenorhabditis elegans*. Histochem J 28: 201–207

Arata Y, Hirabayashi J, Kasai K (1997) The two lectin domains of the tandem-repeat 32-kDa golectin of the nematode *Caenorhabditis elegans* have different binding proterties. J Biochem 121: 1002–1009

Ashwell G, Horford J (1982) Carbohydrate-specific receptors of the liver. Annu Rev Biochem 51: 531–554

Atoda H, Hyuga M, Morita T (1991) The primary structure of coagulation factor IX/factor X-binding protein isolated from the venom of *Trimeresurus flavoviridis*. J Biol Chem 266: 14903–14911

Barondes SH (1981) Lectins: Their multiple endogenous cellular functions. Annu Rev Biochem 50: 207–231

Barondes SH (1984) Soluble lectins: A new class of extracellular proteins. Science 223: 1259–1264

Barondes SH, Castronovo V, Cooper DW, Cummings RD, Drickamer K, Feizi T, Gitt MA, Hirabayashi J, Hughes C, Kasai K, Leffler H, Liu FT, Lotan R, Mercurio AM, Monsigny M, Pillai S, Poirier F, Raz A, Rigby PW, Rini JM, Wang JL (1994a) Galectins: a family of animal β-galactoside-binding lectins. Cell 76: 597–598

Barondes SH, Cooper DNW, Gitt MA, Leffler H (1994b) Galectins: structure and function of a large family of animal lectins. J Biol Chem 269: 20807–20810

Benson B, Hawgood S, Schilling J, Clements J, Damm D, Cordell B, White RT (1985) Structure of canine pulmonary surfactant apoprotein: cDNA and complete amino acid sequence. Proc Natl Acad Sci USA 82: 6379–6383

Bourne Y, Rouge P, Cambillau C (1990) X-ray structure of (α-Man(1–3)β-Man(1–4)GlcNAc)-lectin complex at 2.1 Å resolution. J Biol Chem 267: 197–203

Bourne Y, Bolgiano B, Liao DI, Strecker G, Cantau P, Herzberg O, Feizi T, Cambillau C (1994) Crosslinking of mammalian lectin (galectin-1) by complex biantenary saccharides. Nature Struct Biol 1: 863–870

Caron M, Bladier D, Joubert R (1990) Soluble galactoside-binding vertebrate lectins: a protein family with common properties. Int J Biochem 22: 1379–1385

Cherayil BJ, Weiner SJ, Pillai S (1989) The Mac-2 antigen is a galactose-specific lectin that binds IgE. J Exp Med 170: 1959–1972

Chirariotti L, Wells V, Bruni CB, Malluci L (1991) Structure and expression of the negative growth factor mouse β-galactoside-binding protein gene. Biochem Biophys Res Commun 1089: 54–60

Clerch LB, Whitney P, Hass M, Brew K, Miller T, Werner R, Massaro D (1988) Sequence of a full-length cDNA for rat lung β-galactoside-binding protein: primary and secondary structure of the lectin. Biochemistry 27: 692–699

Cooper DNW, Barondes SH (1990) Evidence for export of a muscle lectin from cytosol to extracellular matrix and for novel secretory mechanism. J Cell Biol 110: 1681–1691

Cooper DNW, Boulianne RP, Charlton S, Farrell EM, Sucher A, Lu BC (1997) Fungal galectins, sequence and specificity of two isolectins from *Coprinus cinereus*. J Biol Chem 272: 1514–1521

Curtis BM, Scharowske S, Watson AJ (1992) Sequence and expression of a membrane-associated C-type lectin that exhibits CD4-independent binding of human immunodeficiency virus envelop gp 120. Proc Natl Acad Sci USA 89: 8356–8360

Dagher SF, Wang JL, Patterson RJ (1995) Identification of galectin-3 as a factor in pre-mRNA splicing. Proc Natl Acad Sci USA 92: 1213–1217

De Caro AM, Bonicel JJ, Rouimi P, De Caro JC, Sarles H, Rovery M (1987) Complete amino acid sequence of an immunoreactive form of human pancreatic stone protein isolated from pancreatic juice. Eur J Biochem 168: 201–207

Decker P, Schweer H, Pohlmann R (1982) Identification of formose sugars, presumable prebiotic metabolites, using capillary gas chromatography/gas chromatography-mass spectrometry of n-butoxime trifluoroacetates on OV-225. J Chromatogr 244: 281–291

Derewenda Z, Yariv J, Helliwell JR, Kalb AJ, Dodson EJ, Papiz MZ, Wan T, Campbell J (1989) The structure of the saccharide-binding site of concanavalin A. EMBO J 8: 2189–2193

Do KY, Smith DF, Cummings RD (1990) Lamp-1 in CHO cells is primary carrier of poly-N-lactosamine chains and is bound preperentially by a mammalian S-type lectin. Biochem Biophys Res Commun 173: 1123–1128

Doege K, Fernandez P, Hassell JR, Sasaki M, Yamada Y (1986) Partial cDNA sequence encoding a globular domain at the C-terminus of rat cartilage proteoglycan. J Biol Chem 261: 8108–8111

Doyle RJ (1994) Introduction to lectins and their interactions with microorganisms. In: Doyle RJ, Slifkin M (eds) Lectin-microorganism interactions. Marcel Dekker, New York, pp 1–65

Drickamer K (1981) Complete amino acid sequence of a membrane receptor for glycoproteins. J Biol Chem 256: 5827–5839

Drickamer K (1988) Two distinct classes of carbohydrate-recognition domain in animal lectins. J Biol Chem 263: 9557–9560

Drickamer K (1992) Engineering galactose-binding activity into a C-type mannose-binding protein. Nature 360: 183–186

Drickamer K (1993) Evolution of Ca^{2+}-dependent animal lectins. Prog Nucleic Acid Res Mol Biol 45: 207–232

Drickamer K, Taylor ME (1993) Biology of animal lectins. Annu Rev Cell Biol 9: 237–264

Drickamer K, Mamon JF, Binns G, Leung JO (1984) Primary structure of the rat liver asialoglycoprotein receptor. Structural evidence for multiple polypepride species. J Biol Chem 259: 770–778

Drickamer K, Dordal MS, Reynolds L (1986) Mannose-binding proteins isolated from rat liver contain carbohydrate-recognition domains linked to collagenous tails. J Biol Chem 261: 6878–6887

Emsley J, White HE, O'Hara BP, Oliva G, Srinivasan N, Tickle IJ, Blundell TL, Pepys MB, Wood SP (1994) Structure of pentameric human serum amyloid P component. Nature 367: 338–345

Ewart KV, Rubinsky B, Fletcher GL (1992) Structural and functional similarity between fish antifreeze protein and calcium-dependent lectin. Biochem Biophys Res Commun 185: 335–340

Fuhlendorff J, Clemmensen I, Magnusson S (1987) Primary structure of tetranectin, a plasminogen Kringl 4-binding plasma protein: homology with asialoglycoprotein receptor and cartilage proteoglycan core protein. Biochemistry 26: 6757–6764

Giga Y, Ikai A, Takahashi K (1987) The complete amino acid sequence of echinoidin, a lectin from the coelomic fluid of the sea urchin *Anthocidaris crassispina*. J Biol Chem 262: 6197–6203

Gitt MA, Barondes SH (1986) Evidence that a human soluble β-galactoside-binding lectin is encoded by a family of genes. Proc Natl Acad Sci USA 83: 7603–7607

Gitt MA, Barondes SH (1991) Genomic sequence and organization of two members of a human lectin gene family. Biochemistry 30: 82–89

Gitt MA, Massa SM, Leffler H, Barondes SH (1992) Isolation and expression of a gene encoding L-14-II, a new human soluble lactose-binding lectin. J Biol Chem 267: 10601–10606

Gitt AM, Wiser MF, Leffler H, Herrmann J, Xia YR, Massa SM, Cooper DNW, Lusis AJ, Barondes SH (1995) Sequence and mapping of galectin-5, a β-galactoside-binding lectin, found in rat erythrocytes. J Biol Chem 270: 5032–5038

Gitt MA, Jordan ET, Leffler H (1997) Galectin-2, galectins-5 and 9, and galectins-4 and 6. Trends Glycosci Glycotech 9: 87–93

Goldstein IJ, Poretz RD (1986) Isolation, physicochemical characterization, and carbohydrate specificity of lectins. In: Liener I, Sharon N, Goldstein IJ (eds) The lectins: properties, functions, and applications in biology and medicine. Academic Press, Orlando, pp 33–247

Goldstein IJ, Hughes RC, Monsigney M, Osawa T, Sharon N (1980) What should be called a lectin? Nature 285: 66

Gritzmacher CA, Mehl VS, Liu FT (1992) Genomic cloning of the gene for an IgE-binding lectin reveals unusual utilization of 5′ untranslated regions. Biochemistry 31: 9533–9538

Gupta D, Brewer CF (1994) Homogeneous aggregation of the 14-kDa β-galactoside specific vertebrate lectin complex with asialofetuin in mixed systems. Biochemistry 33: 5526–5530

Haagsman HP, Haywood S, Aargeant T, Buckley D, White RT, Drickamer K, Benson BJ (1987) The major lung surfactant protein, SP 28–36, is a calcium-dependent carbohydrate-binding protein. J Biol Chem 262: 13877–13800

Hadari YR, Paz K, Dekel R, Mestrovic T, Accili D, Zick Y (1995) Galectin-8. A new rat lectin, related to galectin-4. J Biol Chem 270: 3447–3454

Halberg DF, Proulx G, Doege K, Yamada Y, Drickamer K (1988) A sequnce of the cartilage proteoglycan core protein has lectin-like activity. J Biol Chem 263: 9486–9490

Hirabayashi J (1993) A general comparison of two major families of animal lectins. Trends Glycobiol Glycotechnol 5: 251–270

Hirabayashi J (1994) Two distinct families of animal lectins: speculations on their raisons d'etre. In: van Driessche E, Fischer J, Beeckmans S, Bog-Hansen TC (eds) Lectins, biology, biochemistry, clinical biochemistry, vol 10. Textop, Hellerup, Denmark, pp 205–219

Hirabayashi J (1996) On the origin of elementary hexoses. Q Rev Biol 71: 365–380

Hirabayashi J, Kasai K (1984) Human placenta β-galactoside-binding lectin. Purification and some properties. Biochem Biophys Res Commun 122: 938–944

Hirabayashi J, Kasai K (1988) Complete amino acid sequence of a β-galactoside-binding lectin from human placenta. J Biochem 104: 1–4

Hirabayashi J, Kasai K (1989) Cloning and nucleotide sequence of a full-length cDNA for human 14-kDa β-galactoside-binding lectin. Biochim Biophys Acta 1008: 85–91

Hirabayashi J, Kasai K (1991) Effect of amino acid substitution by site-directed mutagenesis on the carbohydrate recognition and stability of human 14-kDa β-galactoside-binding lectin. J Biol Chem 266: 23648–23653

Hirabayashi J, Kasai K (1993) The family of metazoan metal-independent β-galactoside-binding lectins; structure, function and molecular evolution. Glycobiology 3: 297–304

Hirabayashi J, Kasai K (1994) Further evidence by site-directed mutagenesis that conserved hydrophilic residues form a carbohydrate-binding site of human galectin-1. Glycoconjugate J 11: 437–442

Hirabayashi J, Kawasaki H, Suzuki K, Kasai K (1987) Complete amino acid sequence of 14 kDa β-galactoside-binding lectin of chick embryo. J Biochem 101: 775–787
Hirabayashi J, Ayaki H, Soma G, Kasai K (1989) Production and purification of a recombinant human 14-kDa β-galactoside-binding lectin. FEBS Lett 250: 161–165
Hirabayashi J, Kusunoki T, Kasai K (1991) Complete primary structure of a galactose-specific lectin from the venom of the rattlesnake *Crotalus atrox*. Homologies with Ca-dependent-type lectin. J Biol Chem 266: 2320–2326
Hirabayashi J, Satoh M, Kasai K (1992) Evidence that *Caenorhabditis elegans* 32-kDa β-galactoside-binding protein is homologous to vertebrate β-galactoside-binding lectins. J Biol Chem 267: 15485–15490
Hirabayashi J, Ubukata M, Kasai K (1996) Purification and molecular characterization of a novel 16-kDa galectin from the nematode *Caenorhabditis elegans*. J Biol Chem 271: 2497–2505
Ho SC, Malek-Hedayat S, Wang JL, Scindler M (1986) Endogenous lectins from cultured soybean cells: isolation of a protein immunologically cross-reactive with seed soybean agglutinin and analysis of its role in binding of *Rhizobium japonicum*. J Cell Biol 103: 1043–1054
Hoyle GW, Hill RL (1988) Molecular cloning and sequencing of a cDNA for a carbohydrate-binding receptor unique to rat Kupffer cells. J Biol Chem 263: 7487–7492
Ii M, Kurata H, Itoh N, Yamashina I, Kawasaki T (1990) Molecular cloning and sequence analysis of cDNA encoding the macrophage lectin specific for galactose and N-acetylgalactosamine. J Biol Chem 265: 11295–11298
Inoue S, Kogaki H, Ikeda K, Samejima Y, Omori-Satoh T (1991) Amino acid sequences of two subunits of a phospholipase A2 inhibitor from the blood plasma of *Trimeresurus flavoviridis*. J Biol Chem 266: 1001–1007
Iobst ST, Wormald MR, William IW, Dwek RA, Drickamer K (1994) Binding of sugar ligands to Ca-dependent animal lectins. J Biol Chem 269: 155505–155511
Jackman RW, Beeler DL, Fritze L, Soff G, Rosenberg RD (1987) Human thrombomodulin gene is intron depleted: nucleic acid sequence of the cDNA and gene predict protein structure and suggest sites of regulatory control. Proc Natl Acad Sci USA 84: 6425–6429
Jia S, Wang JL (1988) Carbohydrate-binding protein 35: complementary DNA sequence reveals homology with proteins of the heterogeneous nuclear RNP. J Biol Chem 263: 6009–6011
Kasai K (1990) Biochemical properties of vertebrate 14K beta-galactoside-binding lectins. In: Franz H, Kasai K, Kocourek J, Olsnes S, Shannon LM (eds) Advances in lectin research, vol 3. VEB Verlag Volk und Gesundheit, Berlin, pp 10–35
Kasai K, Hirabayashi J (1996) Galectins: a family of animal lectins that decipher glycocodes. J Biochem 119: 1–8
Kimura M (1983) The neutral theory of molecular evolution, Cambridge University Press, Cambridge
Klion AD, Donelson JE (1994) OvGalBP, a filarial antigen with homology to vertebrate galactoside-binding lectin. Mol Biochem Parasitol 65: 305–315
Kojima K, Ogawa H, Seno N, Yamamaoto K, Irimura T, Osawa T, Matsumoto I (1992) Carbohydrate-binding proteins in bovine kidney have consensus amino-acid sequences of annexin family proteins. J Biol Chem 267: 20536
Kornfeld S (1986) Trafficking of lysozomal enzymes in normal and disease states. J Clin Invest 77: 1–6
Krusius T, Gehlsen KR, Rouslahti E (1987) A fibroblast chondroitin sulfate proteoglycan core protein contains lectin-like and growth factor-like sequence. J Biol Chem 262: 13120–13125
Kubo T, Natori S (1987) Purification and some properties of a lectin from the hemolymph of *Periplaneta americana* (American cockroach). Eur J Biochem 168: 75–82
Lasky LA, Singer MS, Yednock TA, Dowbenko D, Fennie C, Rodriguez H, Nguyen T, Stachel S, Rosen SD (1989) Cloning of a lymphocyte homing receptor reveals a lectin domain. Cell 56: 1045–1055

Lee RT, Lin P, Lee YC (1984) New synthetic cluster ligands for galactose/N-acetylglucosamine-specific lectin of mammalian liver. Biochemistry 23: 4255–4261

Lee RT, Ichikawa Y, Fay M, Drickamer K, Shao MC, Lee YC (1989a) Ligand-binding characteristics of rat serum-type mannose-binding protein (MBP-A). J Biol Chem 266: 4810–4815

Lee RT, Rice KG, Rao BNN, Ichikawa Y, Barthel T, Piskarev V, Lee YC (1989b) Binding characteristics of N-acetylglucosamine-specific lectin of isolated chicken hepatocytes. Biochemistry 28: 8351–8358

Lee RT, Ichikawa Y, Allen HJ, Lee CY (1990) Binding characteristics of galactoside-binding lectin (galaptin) from human spleen. J Biol Chem 265: 7868–7871

Lee Y-M, Leiby KR, Allar J, Paris K, Lerch B, Okarma TB (1991) Primary structure of bovine conglutinin, a member of the C-type animal lectin family. J Biol Chem 266: 2715–2723

Leffler H, Barondes SH (1986) Specificity of binding of three soluble rat lung lectins to substituted and unsubstituted mammalian β-galactosides. Sugar binding specificity. J Biol Chem 261: 10119–10126

Leffler H, Masiarz FR, Barondes SH (1989) Soluble lactose-binding vertebrate lectins: a growing family. Biochemistry 28: 9222–9229

Lennartz MR, Cole FS, Shepherd VL, Wilemen TE, Stahl PD (1987) Isolation and characterization of a mannose-specific endocytosis receptor from human placenta. J Biol Chem 262: 9942–9944

Leung JO, Holland EC, Drickamer K (1984) Characterization of the genes encoding the major rat liver asialoglycoprotein receptor. J Biol Chem 260: 12523–12527

Levi G, Teichberg VI (1981) Isolation and physicochemical characterization of electrolectin, a β-galactoside-binding lectin from the elctric organ of *Electrophorus electricus.* J Biol Chem 256: 5735–5740

Liao DI, Kapadia G, Ahmed H, Vasta GR, Herzberg O (1994) Structure of S-lectin, a developmentally regulated vertebrate β-galactoside-binding protein. Proc Natl Acad Sci USA 91: 1428–1432

Lobsanov YD, Gitt MA, Leffler H, Barondes SH, Rini JM (1993) X-ray crystal structure of the human dimeric S-lac lectin, L014-II, in complex with lactose at 2.9 Å resolution. J Biol Chem 268: 27034–27038

Lodish HF, Kong N, Wikstrom L (1992) Calcium is required for folding newly made subunits of the asialoglycoprotein receptor with the endoplasmic reticulum. J Biol Chem 267: 12753–12760

Madsen P, Rasmussen HH, Flint T, Gromov P, Kruse TA, Honore B, Vorum H, Celis JE (1995) Cloning, expression, and chromosome mapping of human galectin-7. J Biol Chem 270: 5823–5829

Magnaldo T, Bernerd F, Darmon M (1995) Galectin-7, a human 14-kDa S-lectin, specifically expressed in keratinocytes and sensitive to retinoic acid. Dev Biol 168: 259–271

Mandal D, Brewer CF (1992) Cross-linking activity of the 14-kilodalton β-galactoside-specific vertebrate lectin with asialofetuin: comparison with several galactose-spcific plant lectins. Biochemistry 31: 8465–8472

Marschal P, Herrmann J, Leffler H, Barondes SH, Cooper DNW (1993) Sequence and specificity of a soluble lactose-binding lectins from *Xenopus laevis* skin. J Biol Chem 267: 12942–12949

Merkle RK, Zhou Q, Schultz TK, Harper WB, Cummings RD (1989) characterization of S-type lectin purified from porcine heart. Arch Biochem Biophys 274: 404–416

Miller SL (1987) Which organic compounds could have occurred on the prebiotic earth? Cold Spring Harbor Symp Quant Biol 52: 17–27

Miyamura K, Reid KBM, Holmskov U (1994) The collectins – mammalian lectins containing collagen-like regions. Trends Glycosci Glycotechnol 6: 286–309

Muramoto K, Kamiya H (1986) The amino acid sequence of a lectin of the acorn barnacle *Megabalanus rosa.* Biochim Biophys Acta 874: 285–295

Ng NF, Trinh KY, Hew CL (1986) Structure of an antifreeze polypeptide precursor from the sea raven, *Hemitripterus americans.* J Biol Chem 261: 15690–15695

Nurminskaya M, Linsenmayer TF (1996) Identification and characterization of up-regulated genes during chondrocyte hypertrophy. Dev Dyn 206: 260–271

Oda Y, Kasai K (1984) Photochemical cross-linking of β-galactoside-binding lectin to polylactosamino-proteoglycan of chick embryo skin. Biochem Biophys Res Commun 123: 1215–1220

Oda Y, Leffler H, Sakakura Y, Kasai K, Barondes SH (1991) Human breast carcinoma cDNA encoding a galactose-binding lectin homologous to mouse Mac-2 antigen. Gene 99: 279–283

Oda Y, Herrmann J, Gitt MA, Turck CW, Burlingame AL, Barondes SH, Leffler H (1993) Soluble lactose-binding lectin from rat intestine with two different carbohydrate-binding domains in the peptide chain. J Biol Chem 268: 5929–5939

Ohyama Y, Kasai K (1988) Isolation and characterization of the chick 14K β-galactoside-binding lectin gene. J Biochem 104: 173–177

Ohyama Y, Hirabayashi J, Oda Y, Oono S, Kawasaki H, Suzuki K, Kasai K (1986) Nucleotide sequence of chick 14K β-galactoside-binding lectin mRNA. Biochem Biophys Res Commun 134: 51–56

Ozeki Y, Matsui T, Nittal K, Kawauchi H, Takayanagi Y, Titani K (1991) Purification and characterization of β-galactoside-binding lectins from frog (*Rana catesbiana*). Biochem Biophys Res Commun 178: 407–413

Ozeki Y, Matsui T, Yamamoto Y, Funahashi M, Hamako J, Titani K (1995) Tissue fibronectin is an endogenous ligand for galectin-1. Glycobiology 5: 255–261

Paroutaud P, Levi G, Teichberg VL, Strosberg AD (1987) Extensive amino acid sequence homologies between animal lectins. Proc Natl Acad Sci USA 84: 6345–6348

Perillo NL, Pace KE, Seihamer JJ, Baum LG (1995) Apoptosis of T cells mediated by galectin-1. Nature 378: 736–739

Pfeifer K, Hassemann M, Gamulin V, Bretting H, Fahrenholz, Müller WEG (1993) S-type lectins occur also in invertebrates: high conservation of the carbohydrate recognition domain in the lectin genes from the marine sponge *Geodia cydonium*. Glycobiology 3: 179–184

Poirrier F, Robertson EJ (1993) Normal development of mice carrying a null mutation in the gene encoding the L14 S-type lectin. Development 119: 1229–1236

Powell LD, Varki A (1995) I-type lectins. J Biol Chem 270: 14243–14246

Raz A, Lotan R (1987) Endogenous galactoside-binding lectins: a new class of functional tumor cell surface molecules related to metastasis. Cancer Metast Rev 6: 433–452

Richardson (1981) The anatomy and toxonomy of protein structure. Adv Prot Chem 34: 167–339

Sahin V, Tereci O, Schmitt H, Cochlovius B, Johannes T, Schmits R, Stenner F, Luo G, Schobert I, Pfreundschuh M (1995) Human neoplasms elicit multiple specific immune responces in the autologous host. Proc Natl Acad Sci USA 92: 11810–11813

Sai S, Tanaka T, Kosher RA, Tanzer ML (1986) Cloning and sequence analysis of a partial cDNA for chicken cartilage proteoglycan core protein. Proc Natl Acad Sci USA 83: 5081–5085

Sakakura Y, Hirabayashi J, Oda Y, Ohyama Y, Kasai K (1990) Structure of chicken 16-kDa β-galactoside-binding lectin: complete amino acid sequence, cloning of cDNA, and production. J Biol Chem 265: 21573–21579

Sato M, Kawakami K, Osawa T, Toyoshima S (1992) Molecular cloning and expression of cDNA encoding a galactose/N-acetylgalactosamine-specific lectin on mouse tumoricidal macrophages. J Biochem 111: 331–336

Satoh S, Hughes RC (1992) Binding specificity of a baby hamster kidney lectin for H type I and II chains, polylactosamine glycans, and appropriately glycosylated forms of laminin and fibronectin. J Biol Chem 267: 6983–6990

Schweinle JE, Nishiyasu M, Ding TQ, Sastry K, Gillies SD, Ezekowitz AB (1993) Trancated forms of mannose-binding protein multimerize and bind to mannose-rich *Salmonella montevideo* but fail to activate complement in vitro. J Biol Chem 268: 364–370

Sharon N (1987) Bacterial lectins, cell-cell recognition and infectious disease. FEBS Lett 217: 147–157

Siegelman MH, Van de Rijn M, Weissman IL (1989) Mouse lymph node homing receptor cDNA clone encodes a glycoprotein revealing tandem interaction domains. Science 243: 1165–1172

Speck JC Jr (1958) The Lobry de Bruyn-Alberta van Ekenstein transformation. Adv Carbohydr Chem 13: 63–103

Stanek J, Cerby M, Kocourek J, Pacak J (1963) Total synthesis of the monosaccharide series. The monosaccharides. Academic Press, New York, pp 169–174

Suzuki T, Takagi T, Furukohri T, Kawamura K, Nakauchi M (1990) A calcium-dependent galactose-binding lectin from the tunicate *Polyandrocarpa misakiensis*. J Biol Chem 265: 1274–1281

Takahashi H, Komanao H, Kawaguchi N, Kitamura N, Nakanishi S, Natori S (1985) Cloning and sequencing of cDNA of *Sarcophaga peregrina* humoral lectin induced on injury of the body wall. J Biol Chem 260: 12228–12233

Takeya H, Nishida S, Miyata T, Kawada S, Saisaka Y, Morita T, Iwanaga S (1992) Coagulation factor X activating enzyme from Russell's viper venom. J Biol Chem 267: 14109–14117

Terazono K, Yamamoto H, Takasawa S, Shiga K, Yonemura Y, Tochino Y, Okamoto H (1988) A novel gene activated in regenerating islets. J Biol Chem 263: 2111–2114

Tracey BM, Feizi T, Abbott WM, Carruthers RA, Green BN, Lawson AM (1992) Subunit molecular mass assignment of 14 654 Da to the soluble β-galactoside-binding lectin from bovine heart muscle and demonstration of intramolecular disulfide bonding associated with oxidative inactivation. J Biol Chem 267: 10342–10347

Usami Y, Fujimura Y, Suzuki M, Ozeki Y, Nishio K, Fukukki H, Titani K (1993) Primary structure of two-chain botrocetin, a von Willecbrand factor modulator purified from the venom of *Bothrop jararaca*. Proc Natl Acad Sci USA 90: 928–932

Wassarman PM (1987) The biology and chemistry of fertilization. Science 235: 553–560

Weis WI, Drickamer K, Hendrickson WA (1992) Structure of a C-type mannose-binding protein complexed with an oligosaccharide. Nature 360: 127–134

Wells V, Mallucci L (1991) Identification of an autocrine negative growth factor: mouse β-galactoside-binding protein is a cytostatic factor and cell growth regulator. Cell 64: 91–97

White RT, Damm D, Miller J, Spratt K, Schilling J, Hawgood S, Benson B, Cordel B (1985) Isolation and characterization of the human pulmonary surfactant apoprotein gene. Nature 317: 361–363

Whitney PLK, Powell JT, Sanford GL (1986) Oxidation and chemical modification of lung β-galactoside-specific lectin. Biochem J 238: 683–689

Wilson JG, Firth MN, Powell JT, Harrison FL (1989) The sequence of the mouse 14 kDa β-galactoside-binding lectin and evidence for its synthesis on free cytoplasmic ribosomes. Biochem J 261: 847–852

Yamaoka K, Ohno S, Kawasaki H, Suzuki K (1991) Overexpression of a β-galactoside-binding protein causes transformation of BALB3T3 fibroblast cells. Biochem Biophys Res Commun 179: 272–279

Zhou Q, Cummings RD (1993) L14 lectin recognition of laminin and its promotion of in vitro cell adhesion. Arch Biochem Biophys 300: 6–17

Molecular Phylogeny of Eumetazoa: Genes in Sponges (Porifera) Give Evidence for Monophyly of Animals[1]

W.E.G. Müller[2,3]

1 Introduction

At present, the universal phylogenetic tree exhibits a tripartite division of the living world and includes Bacteria ("eubacterial"), Archaea ("archebacterial"), and Eucarya ("eukaryotic") branches (Woese et al. 1991) with the Progenote as the common ancestor (Woese 1987). It is difficult to assess the timely appearance of the first living organism. Some first decipherable evidence of early life has apparently been identified in Swaziland (South Africa) and Pilbara Supergroup (Western Australia) sediments, and dates back 3550 to 3000 Ma (million years) ago; these are stromatolites (see Walter 1996), microfossils, and particulate organic matter (reviewed in Schopf 1994). Eucarya are roughly subdivided further into Diplomonada, Microsporidia, slime molds, ciliates, fungi, plants, and animals (reviewed in Iwabe et al. 1989). Animals are grouped into unicellular Eukaryota, which may have arisen ≈1400 Ma ago (Schopf et al. 1973), and Metazoa, which probably evolved 1000 to 700 Ma ago (reviewed in Willmer 1994).

Metazoa are multicellular and heterotrophic organisms. Their germ cells are formed by meiosis; the fertilization is anisogamous. Sperm and egg form a zygote, which develops into a diploid blastula and usually into a gastrula. The somatic cells of animals are able to differentiate and to form tissues (Hyman 1951). By definition, tissues are associations of uniformly differentiated cells which have lost their omni- or pluripotency and are specialized to carry out specific functions in the organisms. These units interact either with other cell associations or with the extracellular matrix.

Since the original attempts to classify the evolution of living beings by Haeckel (1868), separation into Plantae, Protozoa, Plantozoa (including Porifera) and Animalia-Metazoa, the position of one phylum, the Porifera

---

[1] This review is dedicated to Professor Dr. G. Uhlenbruck (Universität Köln), a pioneer in Invertebrate Immunology

[2] Institut für Physiologische Chemie, Johannes Gutenberg-Universität, Abteilung Angewandte Molekularbiologie, Duesbergweg 6, 55099 Mainz, Germany

[3] Akademie gemeinnütziger Wissenschaften zu Erfurt, Dacherödensches Haus, 99015 Erfurt, Germany

Progress in Molecular and Subcellular Biology, Vol. 19
W.E.G. Müller (Ed.)

(sponges) – multicellular animals – remained uncertain (Campbell 1996). Haeckel (1866) placed the sponges with the Protozoa; later, they were considered to be multicellular animals but separated from the Eumetazoa; mainly due to their simple multicellular organization (reviewed in Nielsen 1995), they were placed into a subkingdom Parazoa (Sollas 1884). Also, the less fortunate name Enantiozoa (Delage 1892), which was created to explain sponge- specific embryological observations, cannot be considered as consistent (Hadzi 1963). Until now, a decision as to whether the Metazoa evolved only once, monophyletic origin, or several times, polyphyletic origin, during evolution remains speculative.

It will be the task of the present chapter to demonstrate, based on molecular data, that (1) sponges are Eumetazoa and (2) all eumetazoan phyla are of monophyletic origin. Previous outlines of this view were given earlier (Gamulin et al. 1994, Müller et al. 1994, 1995; Müller 1995). It appears to be justified to eliminate the term Eumetazoa and to use the term Metazoa for the subkingdom of animals (Animalia). A recent study, based on 18S rRNA sequence data, came to the same conclusion (Cavalier-Smith et al. 1996).

2 Earlier Problems

Several suggestions exist as to the phylogenetic relationships among animal phyla. In the past, data from comparative anatomy, embryology, and paleontology have been used to elucidate potential relationships among distantly related groups (Nielsen 1995). Phylogenetic trees based on these approaches remained speculative (Field et al. 1988). One not universally accepted phylogenetic tree for the Metazoa stems from the Choanoflagellata (syn. Craspedomonadida; Protozoa; Valentine 1994). Their colonies and assemblages "resemble" some cell organization units found in sponges; it is even assumed that the choanoflagellates are directly homologous to choanocytes, a special type of sponge cells (Salvini-Plawen 1978).

During the transition from the protozoan to the metazoan stage, animals had to develop an extracellular matrix as found in all metazoan phyla. It is assumed that gastrulation, the formation of an enteron – with two germ layers, the ectoderm and the entoderm – distinguishes Porifera from the other animals, including Cnidaria (Valentine 1994). With the introduction of the mesodermal element a third germ layer appeared that allows the formation of a variety of organs. Later in evolution a blood-vascular system was formed. A further body space was developed, the hemocoel, which provides hydrostatic functions and an enlarged organ to bathe the heart. Further design elements, segmentations, following either an annelidian or arthropodan plan, led the phyla Annelida and Arthropoda to branch off. Oligomerous coeloms are present in Echinodermata and Chordata.

During the past 15 years molecular biological techniques have advanced considerably, allowing us to take nucleotide-sequence data into consideration for the taxonomy of animals. In a first step, phylogenetic information derived from comparisons of RNA sequences which are highly homologous within the metazoan phyla, e.g. rRNA was used to establish evolutionary trees for animals. 5S rRNA proved to be less reliable, due to an inappropriate rate of evolution (Halanych 1991), whereas data from 18S rRNA as well as 28S rRNA molecules allowed phylogenetic inference (Christen et al. 1991). However, this approach revealed conflicting results. Based on sequence data of 18S rRNA alone, both a polyphyletic (Field et al. 1988) and a monophyletic (Smothers et al. 1994) origin of Metazoa have been concluded. The following reasons for this controversy about the rooting have been given: (1) the molecular phylogenetic methods, based on sequence data of rRNA, have reached their limits; (2) hidden paralogy; or (3) lateral gene transfer (Doolittle et al. 1994).

3 Porifera (Sponges)

3.1 Tissue

Paleontologic data suggest that sponges (Porifera) existed already in the early Cambrian (> 600 Ma ago; Mehl et al. 1997). One view considers Porifera simply to be colonies of choanoflagellata which derived from a separate protozoan lineage and evolved independently from the deuterostomes, protostomes and cnidarians (Barnes 1980). In contrast, according to Weissenfels (cited in Mehlhorn 1989), some sponge cells are already organized into epithelium-like tissues which form simple organs or organ-like assemblies. The most prominent examples are the choanocyte chambers, composed of two kinds of epithelium formed by choanocytes (flagellated cells) and cone cells (cells with a double-conical shape which hang into the chamber formed by choanocytes). The function of the choanocyte chambers as organs or organ-like assemblies is to orient the water flow in one direction from the incurrent to the excurrent channel systems.

3.2 Cell Number

Sponges, belonging to the subphylum Demospongiae, are composed of five different cell types (Simpson 1984), which can be subdivided further according to their function. This number is low compared to the number of cell types found in vertebrates, with over 100 types (Valentine et al. 1993). However, one sponge cell type, the archaeocytes, is considered to be totipotent and capable of undergoing any kind of differentiation (Simpson 1984).

3.3 Genome Size

In earlier studies it was postulated that the level of evolution is correlated to DNA complexity, implying that an increase in DNA complexity parallels the number of different genes in a given organism (Britten and Davidson 1969). However, it was later realized that the correlation between genome size (C value) and "advancement" is poor (Cavalier-Smith 1985; Szathmary and Maynard Smith 1995). More reliable than DNA complexitiy as a reflection of the complexity of a given organism is the number of genes (Szathmary and Maynard Smith 1995).

Recently, the genome sizes in the marine sponges *Geodia cydonium* and *Suberites domuncula* was determined (Imsiecke et al. 1995). In both species, the DNA content was found to be 3.5 pg/cell, corresponding to a C value of 1 670 000 kb. The number of chromosomes (*S. domuncula*) is 32 in the diploid state (Imsiecke et al. 1995). In comparison, the C value of human is 3 400 000 kb (Li and Graur 1991).

3.4 Phylogenetic Position

It has been proposed that Porifera, as a single phylum, might be subdivided into two subphyla: (1) Symplasma with the class Hexactinellida; and (2) Cellularia with the classes Calcarea and Demospongiae (Lafay et al. 1992). However, recent sequence data from 28S rRNAs of Porifera revealed unexpected relationships (Lafay et al. 1992) namely a separation of the Cellularia into groups with different correspondence to the phyla Cnidaria and Ctenophora, hence, allowing no conclusive branching order. Also, 18S rRNA sequence analyses did not prove to be suitable to resolve deep branchings in the phylogenetic tree, including Porifera (Rodrigo et al. 1994).

4 Rationale for a Classification of Sponges as Metazoa

Metazoa are multicellular, heterotrophic and – at least during some stages in their life cycle – motile organisms, with specialized gametic cells. In addition to these basic criteria, Metazoa possess distinct tissues different from the gametic tissue (Nielsen 1995). Focusing on molecular aspects, the following prerequisites are typical for Metazoa and must have evolved during the transition from unicellular Eukaryota to Metazoa.

4.1
Molecules Controlling Tissue Formation

During the evolution of unicellular eukaryotes to multicellular animals, complex molecular systems were established which allow a tuned and controlled interaction between cells themselves, and between cells and extracellular molecules. The interactions of cell surface-associated adhesion- and growth-hormone receptors with their corresponding extracellular ligands, e.g., adhesion factors and growth hormones, play fundamental roles in the development of all animals, irrespective of the level of differentiation, from Porifera to Mammalia.

Two types of adhesion systems exist in Metazoa: Cell-cell and cell-matrix systems. In the extracellular matrix, three main classes of macromolecules interact with cells: (1) the glycosaminoglycans, which are usually covalently linked to proteins (proteoglycans); (2) fibrous proteins, e.g. collagen; and (3) adhesion glycoproteins. These structural elements allow two functional types of cells in animals to differentiate, sheet-forming epithelial cells and motile mesenchymal or mesenchymal-like cells (Morris 1993).

So far several key molecules of tissue formation and maintanance of cell layers, as known from higher Metazoa, have been identified in sponges: collagen (Exposito and Garrone 1990), fibronectin (Labat-Robert et al. 1981), integrin receptor (Pancer et al. 1997), and lectins (Pfeifer et al. 1993a; Wagner-Hülsmann et al. 1996; Müller et al. 1997). Hence, according to these data, major molecules required for the formation of the basal lamina (Martin and Timpl 1987) are already present in sponges. In some sponge species even septate junctions are found (Ledger 1975).

4.2
Signal Transduction Molecules

Vectorial transduction cascades are required to convert extracellular information, e.g., contact to a second cell, to the extracellular matrix, or to hormone or hormone-like signals, into intracellular messages to initiate or to block a physiological reaction. The first event in this cascade is the interaction between a ligand and its corresponding membrane-bound receptor.

The receptors involved in ligand-mediated signal transduction are grouped into three types: type I, type II and type III receptors.

4.2.1
Type I Receptors

Binding of the ligand occurs at the extracellular domain, initiating a chain of phosphorylations via a cytoplasmic domain that itself is a kinase. In most cases the enzymatic activity is a tyrosine kinase.

Until recently receptor protein tyrosine kinases (RTK) have been described only in higher invertebrate species as well as in vertebrate species (e.g., fish, electric ray and amphibians; reviewed in Geer et al. 1994). Among invertebrates, RTKs have been isolated from *Drosophila melanogaster* (the epidermal growth factor receptor-related DER, the insulin receptor-related DILR, and the fibroblast growth factor receptor-related DFR1) and from *Caenorhabiditis elegans* (*let*-23, *kin*-15 and *kin*-16). Furthermore, putative RTK genes have been found in the coelenterate *Hydra vulgaris* (reviewed in Geer et al. 1994; Müller and Schäcke 1996). Recently, we have described an RTK gene from the siliceous sponge *G. cydonium* (Schäcke et al. 1994a,b,c).

4.2.2
Type II Receptors

This type is represented by ion channels (reviewed in Haga et al. 1990). One prominent representative of transmitter-gated ion channels is the nicotinic acetylcholine receptor. It is opened transiently by acetylcholine, which is released from the nerve end at a neuromuscular junction. At present, first cloning data from sponges hinting at the presence of anglutamate receptor homologous to those in other metazoan phyla are available (in preparation).

4.2.3
Type III Receptors

Receptors of this type belong to a family of seven-pass transmembrane receptor proteins (reviewed in Haga et al. 1990). After binding to the ligand, the receptor undergoes a conformational change and transmits the "information" to G-proteins (GTP-binding proteins). The coupling of a given receptor with distinct G-proteins results in an association either with the adenylate cyclase (G_s and G_i proteins) or with phospholipase C (G_p), the primary effector system. Subsequently, either cAMP, Ca^{2+} or diacylglycerol and inositol-trisphosphate are formed, which act as second messengers. In turn, the secondary effector systems, either cAMP-dependent protein kinases (protein kinase A) or Ca^{2+}- and/or diacylgycerol-dependent protein kinases (Ser/Thr kinases), are activated.

One element of this signaling pathway, mediated by type III receptors, has been cloned from *G. cydonium*, the Ser/Thr protein kinase (Kruse et al. 1996).

4.3
Transcription Factors

If similar body plans in animals ranging from Mammalia to Porifera exist, it is conceivable that homologous transcription factors are also involved in the regulation of gene expression, resulting in changes in genetic regulation within cell lineages. The view that a common structure of body plans for all Metazoa exists is also supported by fossil records, which hint at an explosive appearance of a body plan that was the basis for a specialization into those plans found in the different phyla (Knoll 1994).

Two main families of transcription factors exist which control development of body plans and organs in Metazoa (Theißen and Saedler 1995), homeobox genes and MADS-box genes. Homeobox genes have been isolated from humans to sponges (see review, this Vol.). It has been shown that the homeobox genes in humans and *Drosophila* which determine the antero-posterior axis in the body are highly conserved and are expressed in the body in the same order as they display in genomic DNA (De Robertis et al. 1991). This colinearity (Valentine et al. 1996) – the genomic organization of *Hox* genes with their expression in the body regions – suggests that the classification of the higher Metazoa into protostomes and deuterostomes is only a didactic auxiliary. In the freshwater sponge *Ephydatia fluviatilis*, homeobox genes have been described simultaneously by two groups (Coutinho et al. 1994; Seimiya et al. 1994).

MADS-box genes code likewise for proteins involved in the control of development and cell differentiation. They have been described in plants, where they function as flower meristem-identity or organ-identity genes, whereas in animals they code for key elements involved in regulating the expression of a number of immediate-early genes (reviewed in Shore and Sharrocks 1995). As pointed out (Theißen and Saedler 1995), the MADS-box genes add further complexity in Metazoa to the gene network based on the homeobox genes. The phylogenetically oldest metazoan MADS gene has been described in the sponge *G. cydonium* (Scheffer et al. 1997a).

4.4
Migration

Sponges are animals without muscle cells (reviewed in Simpson 1984). However, cell biological and molecular data indicate that sponges react to certain transmitters (reviewed in Müller and Müller 1997) and contract Ca^{2+} dependently using myosin type II as filaments (Lorenz et al. 1996).

Even though sponges are distinguished from other Metazoa by the absence of muscle cells, tissue contraction and expansion is one major prerequisite for these animals to maintain their body shape and to obtain nutrients by filtrating the surrounding water. One contractile element was cloned from *G. cydonium*, nonmuscle myosin type II (Lorenz et al. 1996).

4.5
Molecules of Sensory Organs?

Sponges do not possess specialized nerves; however, several experiments indicate that sponges react to light, e.g., in *Tethya crypta* as a stimulus for oscular opening (Reiswig 1971) or in *E. fluviatilis* as a modulator for the hatching of sponges from gemmules (Rasmont 1970).

All visual systems in Metazoa studied so far, from insects, cephalopods, and vertebrates, contain the visual pigment proteins opsins, which trigger, together with the vitamin A-derived chromophore, a G-protein-mediated signaling cascade (Yarfitz and Hurley 1994). A further prominent protein group in oscular tissues are the crystallins, which are highly expressed in the lens (Wistow and Piatigorsky 1988). A cDNA coding for a βγ-crystallin was successfully cloned from *G. cydonium* (Krasko and Müller, 1997).

4.6 Response to the Environment

Sponges are one of the major phyla found on the marine hard-substrate benthos, both with respect to the number of species and in biomass (Sarà and Vacelet 1973). All sponges live in the aquatic environment; adult specimens are sessile filter-feeders able to ingest particles between 5–50 µm in size through the cells of the mesohyl and the pinacoderm, and microparticles (0.3–1 µm) via the cells of the choanocyte chambers. A specimen of 1 kg, e.g., a small *G. cydonium*, filters ≈24 000 l every day (Vogel 1977).

These nutritional characteristics of sponges could only be established after the acquisition of an efficient system for protection against environmental stress (Scheffer et al. 1997b). Three cDNAs coding for major stress-response proteins have been isolated from *G. cydonium*, the heat shock protein HSP70 (Koziol et al. 1996), DnaJ (Koziol et al., in press) and ubiquitin (Pfeifer et al. 1993b). A further stress protein is the Rab GDP-dissociaton inhibitor (Krasko et al. 1997).

4.7 Morphogens

For the precise temporal and spatial distribution of cells within an animal, morphogens are required. These are low molecular weight compounds which are present in the body in different concentrations, location-specific. This gradient can be recognized by the cells and provides them with information about their actual position and their potential destination.

In vertebrates retinoic acid is one probable morphogen. Among sponges it was first found in the marine sponge *G. cydonium* (Biesalski et al. 1992) and later also in the freshwater sponge *Spongilla lacustris* (Imsiecke et al. 1994). In the first model system, retinoic acid modulates the expression of the oncogene *myb*, while in the second one this morphogen interferes with morphogenesis during the development from larvae by forming specific reduction bodies. These data strongly suggest that morphogens are involved in the organization of body plans in sponges.

4.8 Molecules Potentially Involved in Sponge Immunity

Only little progress has been made in the understanding of invertebrate immunology in molecular terms, especially with respect to sponges. As in all metazoan individuals, sponges have also developed mechanisms to distinguish between self and nonself. However, only little is known about natural challenges to self-integrity in sponges (reviewed in Pancer et al. 1996a); most of the information is available from experimental transplantation studies. Smith and Hildemann (1986), in their extensive review grouped sponge alloimmune responses seen in experimental transplantations into the following two major rejection processes: some species form barriers to separate from nonself tissue; e.g., the marine sponge *Axinella verrucosa* (Buscema and van de Vyver 1983) or the freshwater sponge *Ephydatia muelleri* (Mukai 1992), while others react by cytotoxic factors which destroy the transplant, e.g., the marine sponges *Callyspongia diffusa* (Hildemann et al. 1979) or *G. cydonium* (Pfeifer et al. 1992). However, no definite molecule has been identified which can be considered as being responsible for self/nonself responses in sponges.

In a new attempt, the following three strategies have been pursued: (1) checking for polymorphism in the immunoglobulin-like (Ig-like) domains from the receptor tyrosine kinase of *G. cydonium* and analysis of a potential correlation to an allorecognition reaction (Pancer et al. 1996a); (2) analysis of the corresponding sponge proteins and genes, found also in the human histocompatibility complex-HSP70 (Koziol et al. 1996), proteasome molecules (Pancer et al. 1996b), and tumor necrosis factor (Pfeifer et al. 1992); and (3) searching for cDNAs coding for proteins, featuring scavenger receptor cysteine-rich (SRCR) domains and/or short consensus repeats (SCR; Pancer et al. 1997b). These latter domains have frequently been found in those molecules of vertebrates which are involved in host defense.

5 Genes in Sponges: Coding for Metazoan Proteins

5.1 Model Animal

At least since the Early Cambrian (Atdabatian), the major poriferan taxa, Hexactinellida, Demospongiae, and Calcarea, have been existent, as documented by isolated spicules as well as by more or less completely preserved sponge fossils, e.g., *Solactiniella plumata* (Mehl et al. 1997). The oldest complete sponges in the fossil record are from the Early Cambrian, found at the Sansha section in Hunan (China) in both the lower and upper levels of the Niutitang Formation (Steiner et al. 1993). The oldest isolated spicules identified so far were in thin-sectioned material from the Dengying Formation, "Shibantan" Member, Hubei Province (Steiner et al. 1993). These spicules are mainly monaxial spicules, but some are also rather definite "crosses", suggesting the evidence of triaxones from hexactinellids. The "Shibantan" Member in Hubei is of Late Proterozoic age and stratigraphically equivalent to the Ediacara (South Australia), which is famous for its exceptionally well-preserved Vendian fossils. The Ediacara fauna is normally considered to have begun about 600 Ma ago, although new stratigraphic data place the base of the Ediacara fauna at ≈565 Ma and suggest that it ranges up to the Precambrian/Cambrian boundary (Grotzinger et al. 1995).

Hexactinellida are the oldest group of sponges, as documented from the fossil record of Sansha. The earliest evidence of Demospongiae and Calcarea is the presence of isolated spicules in thin sections of rocks from Early Cambrian (Atdabatian) Archaeocyath mounds of the Flinders Ranges (South Australia). The demosponge spicules are triaenes and aster microscleres of the probably already modified steraster type (Reitner and Mehl 1995), a constituent character of the taxon Geodiidae, which is thus the oldest known group within the Demospongiae.

When we started the studies with *G. cydonium*, analyzing the molecules mediating cell-cell and cell-matrix adhesion biochemically (Müller and Zahn 1973; reviewed in Müller 1982), it was not known that this species belongs perhaps to one of the phylogenetically oldest taxa. Hence, the choice to use this species for molecular biological studies, also with the intention to draw evolutionary conclusions, was fortunate. As a consequence, we are now also starting comparative cloning experiments using species from other classes of Porifera.

G. cydonium (Cellularia: Demospongiae: Tetractinomorpha: astrophorida: Geodiidae) has the shape of a brain and can weigh up to 5 kg. Among systematists it has not yet been decided who originally described this species first. Some authors believe that it was Jameson (1811); others attribute the first description to Lendenfeld (1903), who named this species *Geodia mülleri*. Specimens used for our studies were collected near Rovinj (Croatia) either by dredging at a depth of 20–25 m (biotope: mobile bottom – abundant) or by skin diving at a depth between 3–7 m (biotope: hard bottom – rare; Limski Kanal).

5.2 Gene Structure

One gene, coding for the receptor tyrosine kinase (RTK) has been cloned from *G. cydonium* (Gamulin et al. 1997). The putative structure of the sponge RTK gene (Fig. 1A), *GCTK*Ge, shows: (1) the extracellular part, comprising a Pro/Ser/Thr-rich region and two complete immunoglobulin (Ig)-like domains; (2) the transmembrane domain; (3) the juxtamembrane region; and (4) the catalytic TK domain.

Two introns have been found in the RTK gene: the first is located between the two Ig-like domains and the second intron between the second Ig-like domain and the transmembrane region in the extracellular part of *GCTK*Ge.

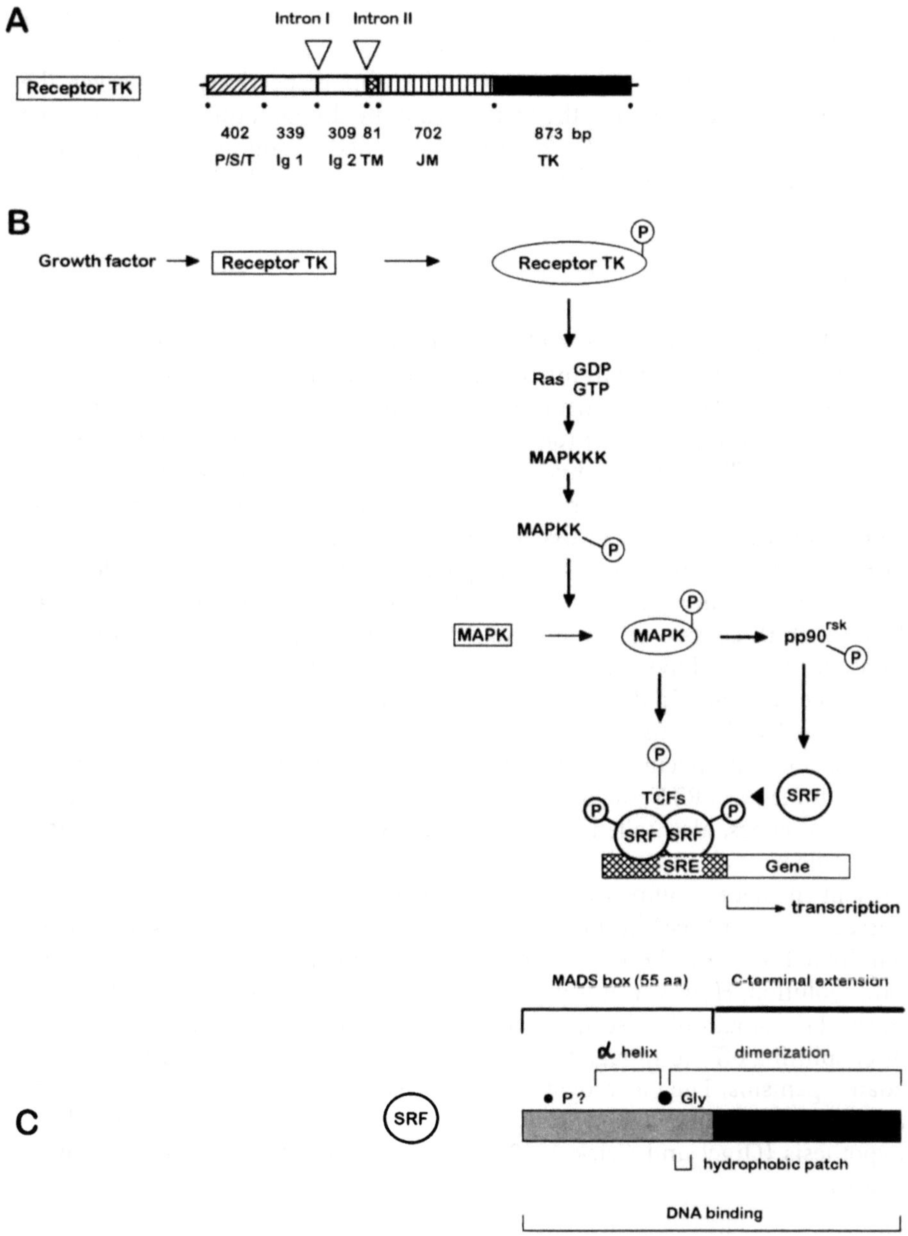

Fig. 1A–C. Postulated signal transduction pathway in *G. cydonium* with the receptor tyrosine kinase (*Receptor TK* or *RTK*) as signal receiver and the serum response factor (*SRF*) as effector. **A** Schematic presentation of the RTK gene from *G. cydonium*. The coding region of the receptor consists of several domains: Pro/Ser/Thr-rich domain, *P/S/T*; immunoglobulin-like (Ig-like) domains 1, *Ig* 1, separated by an intron, *Intron I*, from Ig-like 2, *Ig* 2, then a second intron, *Intron II*, and a transmembrane domain, *TM*; the juxtamembrane region, *JM*; and the TK domain, *TK*. **B** Model for the roles of RTK and SRF in the control of growth factor-initiated gene expression. **C** Scheme of the DNA-binding domain of the SRF with its conserved MADS-box motif and the C-terminal extension which promotes dimerization. Further details about this signaling cascade are given in the text

The introns are of medium size (≈180 and 510 bp, respectively). However, the rest of the gene, comprising the transmembrane domain, the juxtamembrane region and the catalytic TK domain, is encoded by one single exon. Two interesting characteristics of this sponge gene should be highlighted.

5.2.1
Lack of Introns

To the best of our knowledge, all RTK genes studied – with the exception of the *G. cydonium* RTK – contain introns in their TK domain. Some of these genes are more than 100 kb long with over 20 introns, as in the case of the insulin receptor subfamily of RTKs (Seino et al. 1989). The insulin receptor gene, a well-studied RTK, contains five introns in the TK domain. According to the exon theory of genes (Gilbert et al. 1986; Gilbert 1987), it was proposed by Seino and coworkers (1989) that the putative ancestral TK domain may have been assembled from 13 exons; consequently, introns must have been lost in a more or less random fashion from individual genes. Genes encoding human RTKs with three, five or seven Ig-like domains and the TK insert within the catalytic domain have been recently studied in detail (Agnès et al. 1994; Rousset et al. 1995). In these reports introns were found at conserved positions within the TK domains of these genes.

One can only speculate on the structure of the common ancestral gene for all TK domains of RTK, including the *G. cydonium* RTK. According to the exon theory of genes, also called the "introns early" view (Darnell 1978; Doolittle 1978; Gilbert et al. 1986; Gilbert 1987), this common ancestral gene was already present in pieces composed of exons and introns. If this assumption were correct, then (at least) in the sponge *G. cydonium* all introns must have been eliminated from the TK domain during the long period of the sponge's separate evolution. However, it is also – perhaps more – likely that introns, found in the TK domain of existing protein tyrosine kinase genes, were introduced after splitting off the sponge line from the common primitive ancestral metazoan organisms. The most logical way to explain the results obtained from the analysis of the ancient *G. cydonium* RTK gene is to accept the "introns late" hypothesis (Orgel and Crick 1980; Cavalier-Smith 1991; Müller and Müller 1997b).

5.2.2
Intron Polymorphism

Recently, we reported only a relatively low polymorphism among aa sequences across the two Ig-like domains of RTK (Pancer et al. 1996a). However, the intron bridging the two Ig-like domains was found to be highly polymorphic (Pancer and Müller, submitted; Fig. 2). From 4 specimens, 15 independent PCR (polymerase chain reaction) clones were analyzed; 4 clones from sponge specimen no. 3 (termed TKIg3.1-4), 3 from no. 4 (TKIg4.1-3), 5 from no. 5

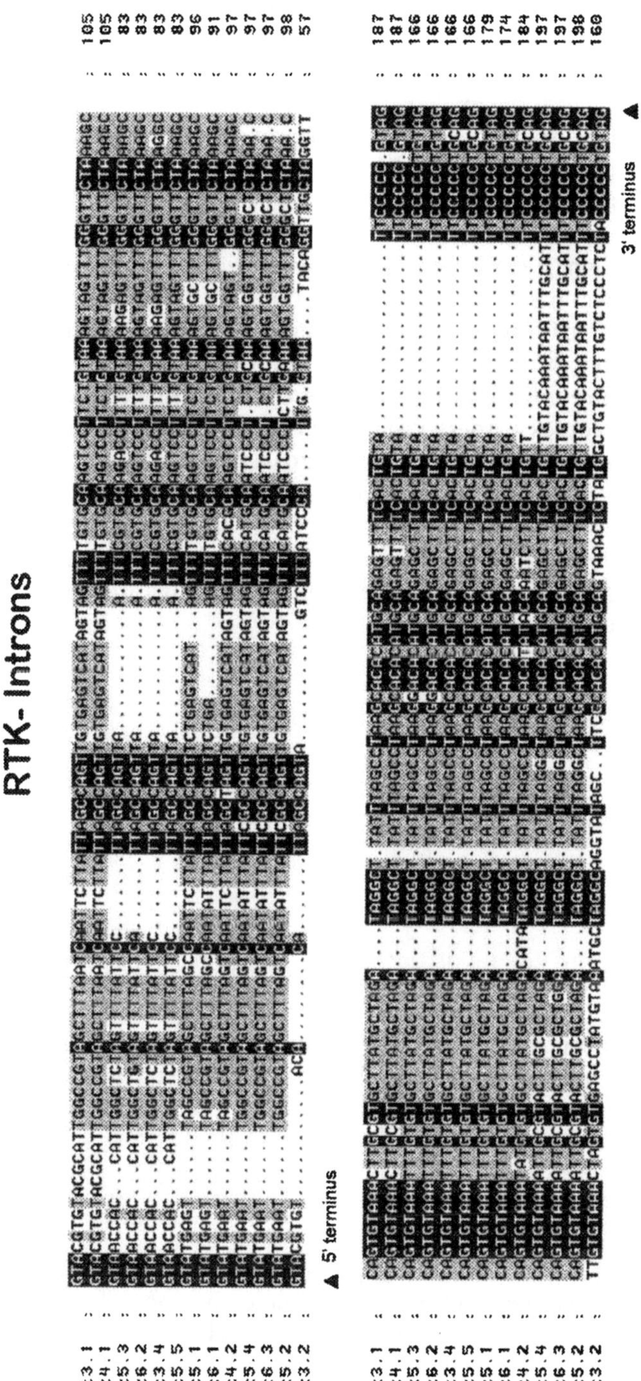

Fig. 2. Introns of the receptor tyrosine kinase (*RTK*) gene; the 5′ and the 3′ *termini* of the introns are marked. Multiple alignment of 13 independent PCR clones obtained by amplification of the first intron located between Ig-like domains 1 and 2 in the GCRTK gene from four specimens of *G. cydonium* are shown: no. 3 (TKInt3.1, 2 and 4), no. 4 (TKInt4.1-2), no. 5 (TKInt5.1-5), and no. 6 (TKInt6.1-3). Nucleotide sequences are arranged by order of similarity; nucleotide positions conserved among all the sequences are shown in *inverted type*, whereas those conserved in at least seven of the sequences are *shaded*; gaps in the alignment are represented by *dots*

(TKIg5.1-5), and 3 from no. 6 (TKIg6.1-3). Sequence analysis of the 15 insert clones revealed only two identical clones from sponge no. 4 (TKIg4.2 and TKIg4.3), and two clones identical among sponges nos. 3 and 4 (TKIg3.1 and TKIg4.1) leaving is sequences to be analyzed.

In contrast to the high polymorphism within the boundaries of the intron (Fig. 2), the putative coding regions and deduced proteins of these 13 genomic PCR clones share pairwise identities of 96–100% at the nt level and 92–100% at the aa level.

These data are probably best explained by the assumption of multiple copies of RTK genes in the genome of *G. cydonium* or other molecules featuring two moderately polymorphic Ig-like domains. However, it remains puzzling as to what evolutionary forces drove such extensive multiplication of an Ig-like bearing gene and the unprecedented divergence of their intronic sequences. Hypothetically, if several copies of these Ig-like genes are already clustered in the genome of one of the most primitive contemporary metazoans and are subjected to sequence divergence in the magnitude presented here, then perhaps such genes were the evolutionary precursors of the vertebrate loci that encode the multicopy V-type and C-type genes which, through V(D)J recombination, rearrange into antibodies and T-cell receptors (Williams and Barclay 1988; Hunkapiller et al. 1989).

5.3 Genes/cDNAs Controlling Tissue Formation

Three cDNAs coding for molecules required for intercellular interactions have been identified in sponges.

5.3.1 *Collagen*

Collagen, a family of proteins widespread in the extracellular matrix of animals, exists already in Porifera (Exposito and Garrone 1990). This first collagen gene in this phylum has been cloned from the freshwater sponge *E. muelleri*. Deduced aa-sequence comparisons indicate that this sponge collagen is homologous to vertebrate and sea urchin collagen.

5.3.2 *Integrin Receptor*

Until now, no direct evidence for the existence of molecules functioning as cell-surface receptors for extracellular matrix molecules has been found in sponges. One major class of these receptors are the integrins. Upon binding ligands, the integrins activate intracellular signaling pathways, e.g., kinases, SH2-SH3 signaling, phospholipid mediators, and matrix assembly (reviewed

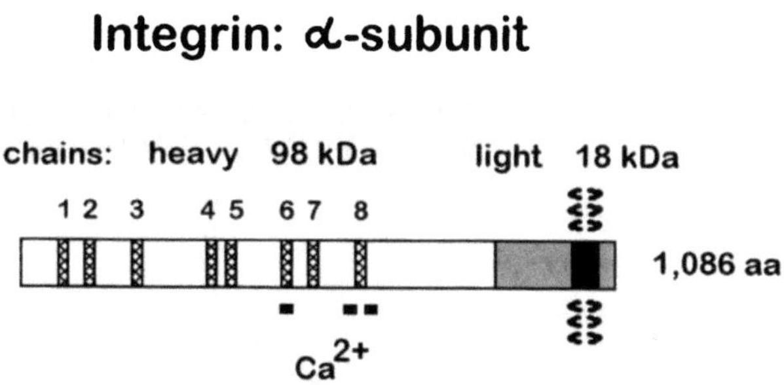

Fig. 3. Schematic presentation of the structural features in *G. cydonium* integrin α-subunit. The heavy and light chains are indicated (the light chain is *shaded*). Positions of eight characteristic repeats of integrins are marked *1* to *8*. Three putative Ca^{2+}-binding sites as well as the C-terminal transmenbrane region are indicated

in Schwartz et al. 1995). From only two invertebrate species have integrin subunits been isolated and cloned: from *C. elegans* and *D. melanogaster* (reviewed in Mehl et al. 1997).

We have isolated and characterized cDNA clones encoding the α subunit of an integrin from the marine sponge *G. cydonium* (Pancer et al. 1997). The open reading frame encodes a 118628kDa polypeptide (Fig. 3). The sequence contains the characteristic consensus pattern for α subunits of integrins, Gly-Phe-Phe-x-Arg, and the characteristic N-terminal amino acids for these molecules. Some evidence suggests that the integrin receptor, very likely present in the sponge plasma membrane, interacts with matrix molecules, e.g., fibronectin (Fig. 4A).

5.3.3 *"C-Type" Lectins-Galectins*

Mammalian/vertebrate lectins are categorized on the basis of their functional and structural characteristics into two main groups (reviewed in: Kasai and Hirabayashi 1996). The "C-type" lectins, which are Ca^{2+}-dependent and have recently been termed galectins, and the "S-type" lectins, which are Ca^{2+}-independent. The "C-type" lectins/galectins contain a long binding domain of 130 aa. Previously, we isolated two major galactose/lactose-specific isolectins (galectins), named lectin-1 and lectin-2, from the extracellular material of *G. cydonium*; these genes were analyzed (Pfeifer et al. 1993a). The typical carbohydrate-binding site of S-type lectins, LH(F)NPR~(G)~V~NT~(G)~W~(T)E~FPF, found in all vertebrate S-type lectins, is present also in the sponge lectin. A homologous lectin was cloned from the nematode *C. elegans* (Hirabayashi and Kasai 1996). A comparison of the carbohydrate-recognition domain of the galectins of invertebrate and vertebrate sequences is shown in Fig. 4B.

A

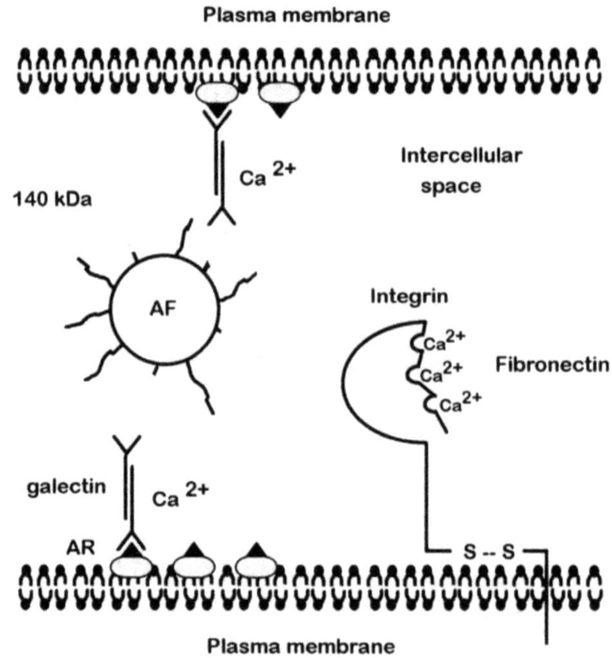

B

```
Sponge LECT-1    42 ALHFNPRYSSTGG--YLVLNTLLNGNWQTEVHPTGFPFPA  79
Sponge LECT-2    43 ·······HA·····--L····S·M··V·AD··R·A·Y···F  80
N32-ii           56 SF·····FDK-----HVIR·S·AANE·GN·EREGKN··EK  90
GBL-EEL          43 ···I···FDAH·DQQAV·V·SFQG···G··QREG····KQ  82
GBL-GALLUS       43 G······FDAH·DVNLI·C·SKKMEE·G··QRE·V···QK  82
GBL-4 MOUSE      43 C······FNAH·DANTI·C··KED·T·G··HREPA···QP  82
GBL-BOVINE       42 C······FNAH·DVNTI·C·SKDA·A·GA·QRESA···QP  81
GBL-RAT          42 C······FNAH·DANTI·C·SKDD·T·G··QRE·A···QP  81
LBL-MOUSE        42 C······FNAH·DANTI·C··KED·T·G··HREPA···QP  81
GBL-HUMAN       156 ·F·····FN-ENNRRVI·C··K·DN··GR·ERQSV···ES 194

Consensus             H-NPRF   ~    V-N   ~  WG-E-R ~ FPF
```

Fig. 4. A Scheme of aggregation factor (*AF*)-mediated cell recognition in *G. cydonium*. A 29-kDa aggregation receptor (*AR*) is inserted into the plasma membrane to which one galectin molecule binds. In the presence of Ca^{2+}, a second galectin molecule binds to the first one. Then these two molecules form a bridge between the AR and the 140-kDa polypeptide, associated with the AF. Following this scheme, the interactions between the AF and the AR involve the galectin which might bind to carbohydrate both at the 140-kDa polypeptide and at the AR. In addition, the sponge contains an integrin receptor (so far only the α-chain has been cloned from *G. cydonium*), which is assumed to interact with fibronectin. **B** Multiple alignment of the conserved region within the galectin-binding domains. The *dots* (•) mark sequence identities between *G. cydonium* galectin-1 and -2 (*Sponge LECt-1 and -2*) and other galectins from *C. elegans* (*N32-ii*), eel (*GBL-EEL*), chicken (*GBL-GALLUS*), mouse (*GBL-4 MOUSE*), bovine (*GBL-BOVINE*), rat (*GBL-RAT*), mouse (*LBL-MOUSE*), and human (*GBL-HUMAN*)

The establishment of a phylogenetic tree for metazoan β-galactose-binding galectins (Hirabayashi and Kasai 1993) indicated that the sponge lectin and the other animal lectins diverged 600–800 Ma ago. Hence, the lectins in Porifera can be considered as ancestral genes for the S-type lectins found in higher metazoan phyla.

5.3.4
Biological Role of Sponge Galectins

The first aggregation factors (AF) were enriched from sponges; especially two species served as sources, the marine sponges, *Microciona prolifera* (Henkart et al. 1973; Weinbaum and Burger 1973) and *G. cydonium* (Müller and Zahn 1973). However, the subsequent elucidation of the detailed function during reaggregation of cells turned out to be difficult. The AF is in fact a complex, as mentioned above, which interacts with a membrane component (AR) that has been identified but not yet purified from *G. cydonium* (Müller et al. 1976). The AR is of lower molecular weight. In a recent study, monoclonal antibodies were used as tools to identify the binding domains of the AF (Wagner-Hülsmann et al. 1996). A 140-kDa polypeptide was found to participate in the reaggregation process. This polypeptide of the AF very likely interacts with the galectin. The galectin itself is the main component which links individual AF molecules to the AR at the plasma membrane and consequently links two cells together (Fig. 4A). The galectin from *G. cydonium* forms in the presence of Ca^{2+} multimers; two and more monovalent galectin molecules associate (Diehl-Seifert et al. 1985; Gramzow et al. 1989).

Immunohistological studies revealed that the spherulous cells contain and perhaps synthesize the galectin. Confocal laser-scanning microscopical analysis demonstrated that both the galectin and the AF are present at the rim of the cells (Wagner-Hülsmann et al. 1996). From this finding we conclude that the sponge galectin is one of the key molecules involved in cell-cell adhesion.

5.4
Signal Transduction Molecules

Two molecules have been cloned from *G. cydonium*: first, the receptor tyrosine kinase (RTK) belonging to the type I receptors, and second, two serine/threonine kinases, members of the type III receptors-initiated signal transduction cascade.

5.4.1
Receptor Tyrosine Kinase

Protein tyrosine kinases [PTKs] represent a large group of enzymes that specifically phosphorylate tyrosine residues (Hardie and Hanks 1995). They play important roles in the response of cells to different extracellular stimuli

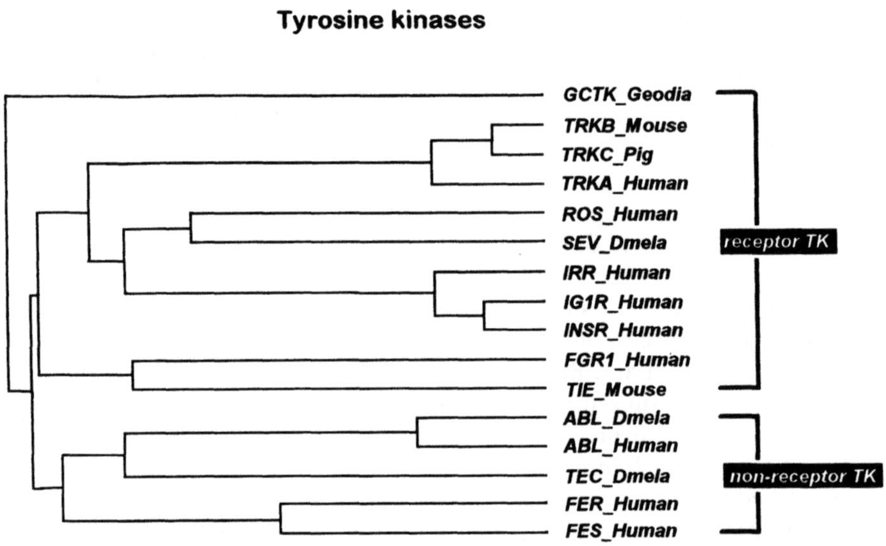

Fig. 5. Dendrogram deduced from the multiple alignment of the TK domain of RTK gene of *G. cydonium* [*GCTK_Geodia*] with those of ten receptor TKs: neurotrophin-4 receptor (*TRKB_Mouse*); NT-3 growth factor precursor (*TRKC_Pig*); nerve growth factor receptor (*TRKA_Human*); ROS proto-oncogene tyosine kinase (*ROS_Human*); *D. melanogaster* sevenless receptor PTK (*SEV_Dmela*), insulin receptor-related receptor (*IRR_Human*), insulin-like growth factor 1 receptor precursor (*IG1R_Human*); insulin receptor precursor (*INSR_Human*); basic fibroblast growth factor receptor 1 precursor (*FGR1_Human*) and TIE protein tyrosine kinase (*TIE-Mouse*); and of five non-receptor TKs DASH/ABL proto-oncogene tyrosine kinase from *D. melanogaster* (*ABL_Dmela*), ABL proto-oncogene tyrosine kinase (*ABL_Human*); *D. melanogaster* SRC protein tyrosine kinase (*TEC_Dmela*), FES/FPS-related PTK (*FER_Human*); and FES/FPS protein-tyrosine kinase (*FES_Human*)

and are essential proteins, most notably for control of growth and differentiation. Many PTKs serve as receptors and signal transducers for circulating peptide hormones and growth factors. PTKs, together with Ser/Thr kinases, represent the largest known protein superfamily.

All PTKs possess a closely related tyrosine kinase TK domain which is specific for the phosphorylation of tyrosine only (Hunter et al. 1992). PTKs are divided into two major groups, the receptor tyrosine kinases (RTKs), which are membrane-spanning molecules with similar overall structural topologies, and the nonreceptor TKs, also composed of structurally similar molecules. Members of the PTKs are further classified on the basis of their structural (functional) similarities and divided into more than twenty subfamilies (Hunter et al. 1992).

As mentioned above, the putative structure of the sponge RTK protein shows: (1) in the extracellular part a Pro/Ser/Thr-rich region, two complete immunoglobulin (Ig)-like domains; (2) the transmembrane domain; (3) the juxtamembrane region; and (4) the catalytic TK domain (Schäcke et al. 1994a–d; Müller and Schäcke 1996; Pancer et al. 1996a; Fig. 1A).

A homology search, with the *G. cydonium* TK domain aa sequence deduced from the gene as well as from the cDNA sequence, was performed. All 50 sequences most homologous to sponge RTK contained the TK domains of PTKs. The dendrogram of the alignment of TK domains of 15 PTKs is shown in Fig. 5. All RTKs fall in one branch of the tree while the non-receptor TKs are grouped in a second one; sponge RTK is placed in a separate branch, which splits off first from the common tree of metazoan PTKs.

According to Kimura (1983), aa substitutions per site, including possible multiple substitutions (*K*aa), can be calculated from the percentage of identity using the equation $Kaa = -\ln \times$ [observed identity (%)/100]. This gives *K*aa values of 1.17–1.20 for the sponge RTK. The standard value for the average rate of nonsynonymous substitutions in DNA has been set at 0.9×10^{-9} per site and year. Using this (average) rate for proteins and the obtained values for *K*aa, it was possible to estimate the time when sponge tyrosine kinase started to diverge from the common ancestral gene of class II TKs ($Kaa/2$: 0.9×10^{-9}); a time of 650–665 Ma was calculated (Schäcke et al. 1994d).

5.4.2
Serine/Threonine Kinases

Protein kinases C (PKCs) comprise closely related Ser/Thr kinases, ubiquitously present in animal tissues; they express their activities in response to second messengers, e.g., Ca^{2+} or diacylglycerol. Two PKCs have been sequenced from *G. cydonium* (Kruse et al. 1996). One sponge PKC, GCPKC1, belongs to the "novel" (Ca^{2+}-independent) PKC (nPKC) subfamily, while the second one, GCPKC2, has the hallmarks of the "conventional" (Ca^{2+}-dependent) PCK (cPKC) subfamily. The alignment of the Ser/Thr catalytic kinase domains, of the predicted aa sequences for these cDNAs with respective segments from previously reported sequences, revealed highest homology to PKCs from animals but also distant relationships to Ser/Thr kinases from protozoa, plants, and bacteria.

However, a comparison of the complete structures of the sponge PKCs, which are – already – identical to those of nPKCs and cPKCs from higher Metazoa, with the structures of protozoan, plant, and bacterial Ser/Thr kinases indicates that the metazoan PKCs must be distinguished from the nonmetazoan enzymes (Fig. 6). These data indicate that metazoan PKCs and nonmetazoan Ser/Thr kinases have a universal common ancestor with respect to the kinase domain, but they differ in the overall structural composition.

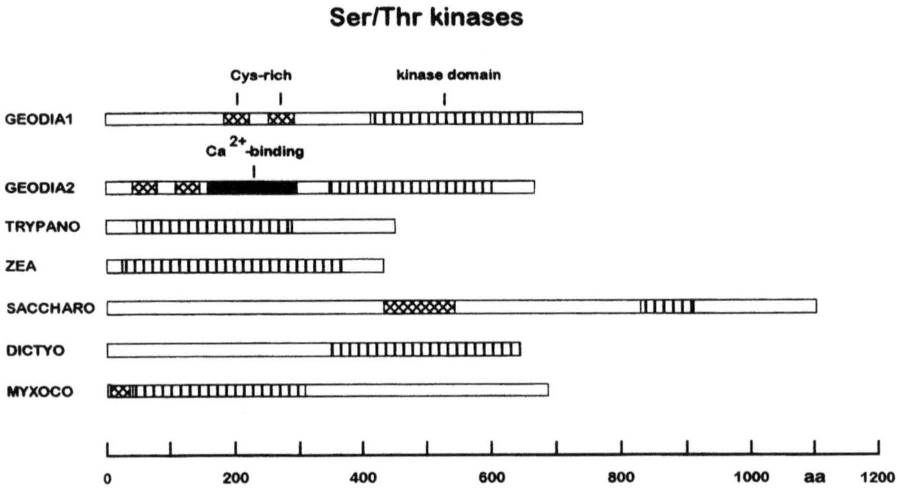

Fig. 6. Comparison of the domain structures of the metazoan PKC, here with the examples of the sponge GCPKC1 and GCPKC2, with the nonmetazoan Ser/Thr kinases. The following sequences have been selected: Metazoa: *G. Cydonium* GCPKC1 and GCPKC2 (*GEODIA1* and *GEODIA2*); Protozoa: *Trypanosoma brucei* (*TRYPANO*); Planta: *Zea mays* (*ZEA*); yeast: *Saccharomyces cerevisiae* (*SACCHARO*); slime mold: *Dictyostelium discoideum* (*DICTYO*), and Prokaryote: *Myxococcus xanthus* (*MYXOCO*). The presentation is according to Hardie and Hanks (1995): in Cys-rich repeats (*cross-hatched*), C2 (Ca^{2+}-binding site; *black*), and kinase domain (*vertically hatched*). The length of the deduced protein is given in a scale of aa numbers

5.5 Transcription Factors

5.5.1 *Homeodomain Proteins*

The nucleotide sequence of the cDNA from the freshwater sponge *E. fluviatilis*, comprising the complete homeodomain sequence, has been identified (Coutinho et al. 1994; Seimiya et al. 1994). The deduced polypeptide of the homeodomain of the sponge shows the characteristic features known from other members of this family. The characteristics of these molecules will be discussed in separate chapters of this volume.

5.5.2 *Serum Response Factor*

Stimulation of receptor tyrosine kinases (RTKs), e.g., the epidermal growth-factor tyrosine kinase receptor, causes the expression of a number of genes without the need for prior protein synthesis (reviewed in Greenberg and Ziff 1984). The rate of transcription of those immediate-early genes, with the proto-oncogene c-*fos* as a prominent member, increases, e.g., in fibroblasts, within 5 min 50-fold and returns to a basal level after 30 min. After stimulation of a cell in response to the binding of a growth factor RTK to its cognate ligand, a series of phosphorylation events are triggered, among them the activation of the mitogen-activated protein (MAP) kinase cascade, which in turn results in the phosphorylation of the ternary complex factors (TCFs) and the ribosomal S6 protein kinase $pp90^{rsk}$ (Sun and Tonks 1994). The $pp90^{rsk}$ kinase itself phosphorylates the serum response factor, SRF (Sun and Tonks 1994; Fig. 1B). After phosphorylation, the SRF dimerizes and displays high affinity to the serum response element (SRE) within the *c-fos* promoter region, where it, together with TCFs, controls *c-fos* expression (reviewed in Sun and Tonks 1994; Shore and Sharrocks 1995).

A cDNA clone encoding the sequence-specific DNA-binding protein, SRF, has been isolated and analyzed from *G. cydonium* (Scheffer et al. 1997a). Sequence comparisons revealed that the aa sequence shares high homology to human and *Xenopus laevis* SRFs. Like these vertebrate sequences, the sponge sequence displays the MADS-box motif in the DNA-binding domain (Fig. 1C). The SRF (and other transcription factors, such as ternary complex factors) recognizes the serum response element (CArG box) in promoter regions of a series of cellular immediate-early genes, whose expression is controlled by growth factors. A related CArG-box sequence is present in the sponge gene, encoding the receptor tyrosine kinase (Gamulin et al. 1997).

5.6 Migration

Sponges contain elements of a neuromuscular system. As examples, acetylcholine and cholinesterase have been identified in sponges (Lentz 1966; Thiney 1972), and a special cell type, the myocytes, which occur in contractile membranes of the water canal system, possess microfilaments (Thiney 1972). However, only recently did the first experimental evidence become available which identifies components controlling migration of sponge cells. One factor is sulfate, which is required for the migration of cells from *M. prolifera* (Kuhns et al. 1995).

We analyzed the role of Ca^{2+} for the migration of sponge cells (Lorenz et al. 1996). From earlier studies it was known that in the sponge-cell system the intracellular Ca^{2+} level increases in response to binding of the homologous aggregation factor (Müller et al. 1987). In turn, Ca^{2+} activates the protein kinase C. It is documented that after addition of ionomycin to cells from the marine sponge *S. domuncula*, suspended in Ca^{2+}-containing seawater, they showed an increase in their motility from 0.2 µm/min (absence of the ionophore) to 3.7 µm/min. As in other metazoan systems, ionomycin induced Ca^{2+} entry into sponge cells. The saturation concentration was around 5 µM. This value is higher than that determined in cells of higher animals, e.g., in endothelial cells (1 µM). Surprising was the finding that ionomycin did not increase the intracellular calcium-ion concentration – $[Ca^{2+}]_i$ – if Ca^{2+} was absent in the extracellular environment of the sponge cells. This finding suggests that sponges have no intracellular Ca^{2+} stores (Lorenz et al. 1996).

These data show that ionomycin increases the motility of sponge cells if they are incubated in Ca^{2+}-containing seawater but not if the cells are suspended in Ca^{2+}-free seawater. Ca^{2+} is known to be one regulator for the dynamic assembly-disassembly of nonmuscle myosin II process (survey: Tan et al. 1992), and the gradient of this ion regulates the sol/gel transition occuring during cell migration. Furthermore, Ca^{2+} acts as a second messenger and activates kinases, which in turn catalyze myosin phosphorylation and ultimately contraction of cells (Allen and Walsh 1994).

Therefore, we looked if myosin is present and acts as a motor for cell migration in sponges. The partial sequence of myosin was identified from the sponge *G. cydonium* (Lorenz et al. 1996). The deduced aa sequence revealed highest homology with nonmuscle myosin type II from chicken, human, gizzard, and *D. melanogaster*. A rooted phylogenetic tree of the myosin tail sequences displaying highest homology to sponge shows that myosin II from *G. cydonium* represents the most ancestral one. The aa sequence shows the characteristic 28-residue repeats known from the tails of other myosin sequences (McLachlan and Karn 1982). This aa arrangement allows coiled coil formation of two-tail α-helices.

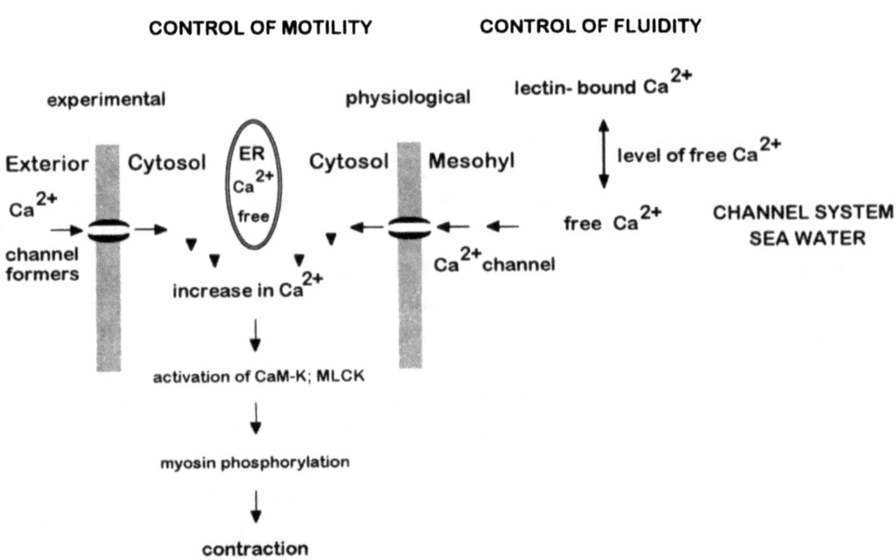

Fig. 7. Proposed model for a Ca^{2+}-regulated contraction and motility in sponge cells. It is suggested that the increased level of $[Ca^{2+}]_i$ causes an activation of myosin via Ca^{2+}-dependent protein kinase, the Ca^{2+}/calmodulin-dependent protein kinase type II (*CaM-K*) and/or the myosin light-chain kinase (*MLCK*). Extracellularly, in the mesohyl compartment, Ca^{2+} is reversibly bound to the lectin; the gradient of this ion regulates the sol/gel consistency in the mesohyl. Further details are given in the text

At present, we assume that the motility of sponge cells is controlled by Ca^{2+}. An increased level of $[Ca^{2+}]_i$ – mediated experimentally by ionomycin – causes an activation of myosin via Ca^{2+}-dependent protein kinase, the Ca^{2+}/calmodulin-dependent protein kinase type II, and/or the myosin light-chain kinase. Under in vivo conditions Ca^{2+} can enter sponge cells via existing Ca^{2+} channels in the plasma membrane (Müller et al. 1996; Fig. 7). Extracellularly, Ca^{2+} is present either in a free state (seawater) or bound to a lectin, as shown for the *G. cydonium* system (Diehl-Seifert et al. 1985). The fluidity in the mesohyl is controlled by lectin. Not associated with Ca^{2+}, the lectin molecules remain separate; while bound to Ca^{2+}, the lectin associates to sheet-like assemblies. Future studies must show if, in sponges, besides myosin II, myosin I is also present, which in concert with other actin-binding proteins regulates the oriented cell migration as in eosinophils (Brundage et al. 1991).

5.7 Molecules of Sensory Organs?

In a first attempt to search for molecules usually found in sensory organs of higher Metazoa we searched for crystallins, one major protein in those organs. Crystallins are categorized into two classes, the ubiquitous crystallins and the taxon-specific crystallins (reviewed in Wistow and Piatigorsky 1988; Jaenicke 1994; Kodama and Eguchi 1994). No structural or functional characteristics are common to all crystallins. The α-, β- and γ-crystallins are classed as ubiquitous crystallins and are found in almost all vertebrate species. The second class, the taxon-specific crystallins, includes a series of "enzyme crystallins", among them δ-, ζ-, and ρ-crystallin, which display catalytic functions. One member of the ubiquitous crystallins, α- crystallin, is structurally related to the ubiquitous small heat-shock proteins (Merck et al. 1993), whereas the βγ-crystallins form a separate class of homologous proteins (Wistow 1993). Until now, no molecular sequence data has been available from βγ-crystallins in invertebrates.

We succeeded in the isolation and characterization of a cDNA, coding for the βγ-crystallin molecule from *G. cydonium* (Krasko and Müller, 1997). The deduced amino-acid sequence consists of 163 aa residues and comprises four repeated motifs, which compose the two domains of the βγ-crystallin (Fig. 8A). Motif 3 contains the characteristic βγ-crystallin "Greek key" motif signature, while in each of the three other repeats, one aa residue is replaced by an aa with the same physicochemical property.

The sponge peptide shows striking homologies to vertebrate βγ-crystallins. The phylogenetic studies were performed by neighbor-joining analysis and revealed that motif 4 of the sponge βγ-crystallin is the common ancestor of not only the sponge motifs 1–3 but also the vertebrate βγ-crystallin motifs. For the phylogenetic tree analysis, the sponge βγ-crystallin sequence as well as the vertebrate sequences were cut into the four motifs and aligned on the basis of

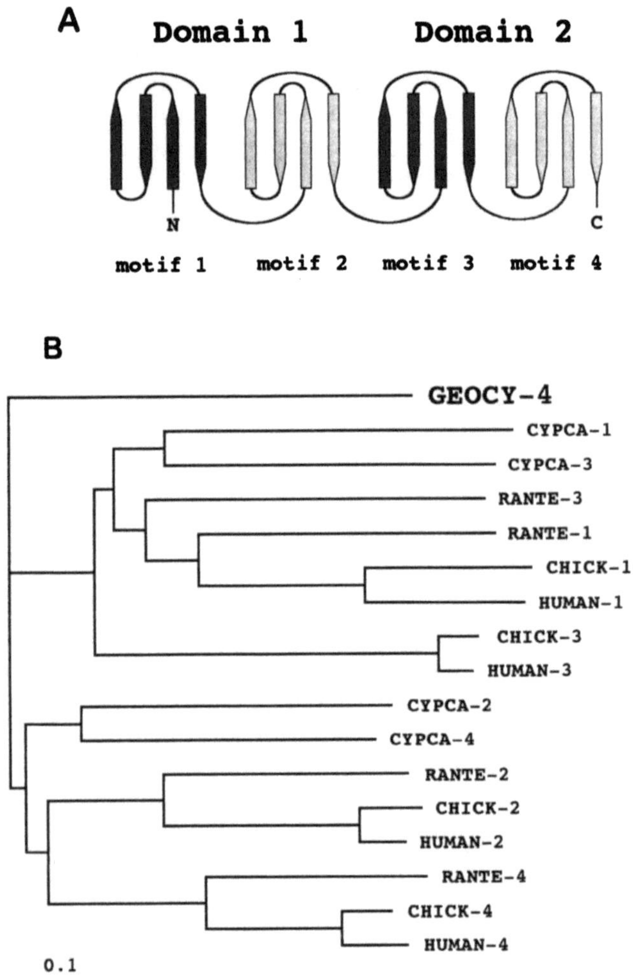

Fig. 8A–B. Sponge βγ-crystallin. The cDNA was cloned from *G. cydonium*. **A** Schematic representation of the βγ-crystallin folding pattern. Two Greek-key motifs form one domain. *Domains 1* and *2* form the monomeric βγ-crystallin. **B** Phylogenetic tree (unrooted) of metazoan βγ-crystallin motifs. The comparison is based on the aa sequence of the four motifs, present in the sequences from human *HUMAN-1 to -4*, chick (*CHICK-1 to -4*), frog (*RANTE-1 to -4*) and fish (*CYPCA-1 to -4*) as well as the motif 4 from the *G. cydonium* sequence *GEOCY-4*. Calculation was performed by neighbor-joining. *Scale bar* indicates an evolutionary distance of 0.1 aa substitutions per position in the sequence

maximum aa similarity. Neighbor-joining and distance-matrix analyses revealed that motif 4 of the *G. cydonium* sequence is the common ancestor for the vertebrate motifs (Fig. 8B). The robustness of the tree can also be deduced from the fact that the tree is trichotomous, with GEOCY-4 forming one branch, *all* motifs 1 and 3 in the second, and *all* motifs 2 and 4 in the third branch. The grouping of motifs 1 and 3 in one class and motifs 2 and 4 in a second is characteristic for vertebrate βγ-crystallins (Wistow and Piatigorsky 1988).

5.8 Response to the Environment

In general, five classic families of stress proteins are found in eukaryotes; four of them are grouped according to their molecular weights as heat-shock proteins HSP90, HSP70, HSP58-60 and HSP20-30, whereas the fifth HSP is termed ubiquitin (Schlesinger et al. 1982). Recently, a new group of stress proteins, DnaJ molecules, have been identified in eukaryotes and termed eDnaJ. A further putative biomarker, present in *G. cydonium*, is the GDP dissociation inhibitor.

5.8.1 *Heat-Shock Protein 70*

The cDNA from the sponge *G. cydonium* has been cloned and applied to determine the expression of HSP70 in reactions of the organism to temperature stress (Koziol et al. 1996). Analysis of the deduced aa sequence from *G. cydonium*, GCHSP70, revealed that it has the typical characteristics of a HSP70 comprising (1) the heat-shock HSP70 protein family signature 1, IDLGTTYS; (2) the heat-shock HSP70 protein family signature 2, DLGGGTFD; and (3) the ATP-binding site with the sequence LxFDxGGGxxD; the latter segment binds the Mg^{2+} ion and the β-phosphate of MgATP. In addition, GCHSP70 displays the following characteristic sites: (1) one ATP/GTP binding-site motif A (P-loop); and (2) two bipartite nuclear targeting sequences. Alignment studies revealed that *G. cydonium* cDNA GCHSP70 belongs to the subfamily of cytoplasmic HSP70s. Exposure of *G. cydonium* to temperature stress resulted in an increased synthesis of mRNA coding for HSP70 (Koziol et al. 1996).

5.8.2 *Ubiquitin*

After having cloned the cDNA for ubiquitin (Pfeifer et al. 1993b), we demonstrated that the ubiquitin pathway exists in sponges. Moreover, the expression of this protein is regulated by cell-adhesion molecules during reaggregation of cells. The characteristics of this molecule will be discussed in a separate chapter of this volume.

As in other organisms the level of ubiquitin in sponges varies in accordance with physiological and pathophysiological conditions (Mayer et al. 1991). Therefore, it was interesting to determine if the ubiquitin protein is expressed differentially within the organism. Surprisingly, it was found that detectable amounts of ubiquitin protein are present only in the cortex (the surface of the sponge exposed to the environment) and not in the medulla region (the internal zone of the sponge), while the level of ubiquitin mRNA is almost identical in both compartments (Pfeifer et al. 1993b).

5.8.3
DnaJ

Recently, eukaryotic homologues of DnaJ have been identified and termed eDnaJ (summarized in Hightower et al. 1994). eDnaJ is present in the nucleus/nucleolus of heat-shocked cells and returns to the cytoplasm during the period of recovery. It has been assumed that eDnaJ is involved in: (1) targeting of HSP70 proteins to a particular cellular compartment; (2) the local accumulation of HSP70; (3) binding of HSP70 to a given substrate; (4) stabilization of HSP70-substrate complex; and (5) stimulation of the dissociation of HSP70 from its substrate.

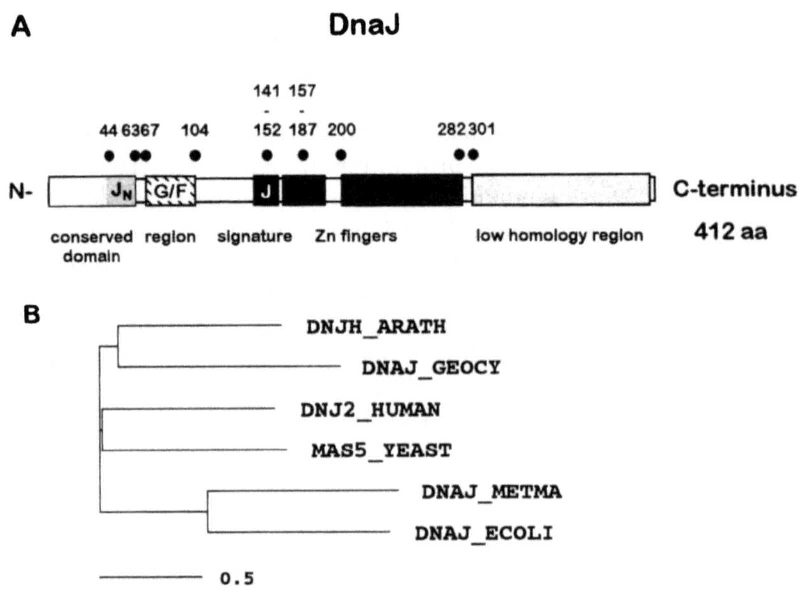

Fig. 9A–B. DnaJ protein of *G. cydonium*. **A** Structure of the putative protein with its N-terminal conserved region, including the J_N domain, the Gly/Phe-rich motif (*G/F*), the further conserved DnaJ domain signature CxxCxGxG{AGS}xxG (J), and two further Zn fingers. The C-terminal region of DnaJ is less conserved. **B** Unrooted phylogenetic tree computed from alignments of the deduced aa sequence of the sponge DnaJ (DNAJ_GEOCY) with the corresponding selective sequences from a metazoan (human cDNA encoding DnaJ protein homologue, DNJ2_HUMAN); yeast (MAS5, a *S. cerevisiae* homologue of DnaJ involved in mitochondrial protein import, MAS5_YEAST); plant (DnaJ protein homologue ATJ from *Arabidopsis thaliana*, DNJH_ARATH); bacterium (*E. coli* DnaJ and GrpE heat-shock proteins jointly stimulate ATPase activity of DnaK, DNAJ_ECOLI); and an archaebacterium (DnaJ in *Methanosarcina mazei*, DNAJ_METMA). Calculation was performed by neighbor-joining analysis. *Scale bar* indicates an evolutionary distance of 0.5 aa substitutions per position in the sequence

The deduced aa sequence of DnaJ_GEOCY is comprised of the motifs characteristic for most DnaJ family members (Bork et al. 1992; Koziol et al., in press; Fig. 9A). It consists of the N-terminal conserved region (Georgopoulos et al. 1994; Tsai and Douglas 1996), including the J_N domain, which is followed by the G/F (Gly/Phe-rich motif) region (Well et al. 1995). Additionally, the further conserved DnaJ domain signature CxxCxGxG[AGS]xxG (Tsai and Douglas 1996) is present in the C-terminal region, which is followed by two Zn fingers. The C-terminal region of DnaJ is less conserved.

Like other heat-shock proteins the DnaJ (DnaJ homologue) proteins are also highly conserved and are found in Eucarya, Bacteria and Archaea. Therefore, an alignment with a subsequent unrooted phylogenetic analysis was performed, including selected sequences from the respective regna and the sponge DnaJ sequence. As expected, the sponge sequence clusters together with the members of eucarya and differs from those of two other regna (Fig. 9B).

5.8.4 *GDP Dissociation Inhibitor*

Eukaryotic cells generate and maintain their organization during growth differentiation and reaction to environmental stressors by regulation of the intracellular membrane traffic. The eukaryotic proteins which are involved in this transport belong to the family of Rab proteins, small Ras-like GTPases (summarized in Boguski and McCormick 1993; Simon and Zerial 1993). Like other G-proteins, Rab proteins undergo cycles of GTP binding, GTP hydrolysis and exchange. Hence, two forms of Rab exist, the active GTP-bound form and the inactive GDP-bound form. Three main factors are involved in this cycle: (1) the GTPase-activating protein(s) (GAPs), which initiate the conversion of the GTP-bound to the GDP-bound form of Rab; (2) the GDP dissociation inhibitor (GDI) that retransports the GDP-associated Rab from the acceptor membrane to the donor membrane; and (3) the guanine-nucleotide-exchange factor (GEF) that replaces GDP with GTP. When GDI is in the nonfunctional form, the intracellular traffic is halted (Takaishi et al. 1995). The mechanism by which Rab proteins function is shown in a scheme (Fig. 10A). It becomes obvious that, under conditions during which the GDP dissociation inhibitor is in the nonfunctional form, the intracellular traffic is halted.

We have isolated the Rab GDI from *G. cydonium* (Krasko et al. 1996). The deduced aa sequence (M_r 50 215) displays high homology to vertebrate, invertebrate, and also yeast GDIs. Northern blot analysis revealed that the 1.6 kb sponge transcript that hybridized with the sponge *GDI* is down-regulated under temperature stess; in contrast, the transcription of the sponge *HSP70* is strongly enhanced (Krasko et al. 1996).

The deduced aa sequence of the sponge *GCGDI* (GDI_GEODIA in Fig. 10B) was aligned with the corresponding sequences from human GDI class β, mouse GDI β, rat GDI β, human GDI α, bovine GDI α, rat GDI α, mouse GDI α, *C.*

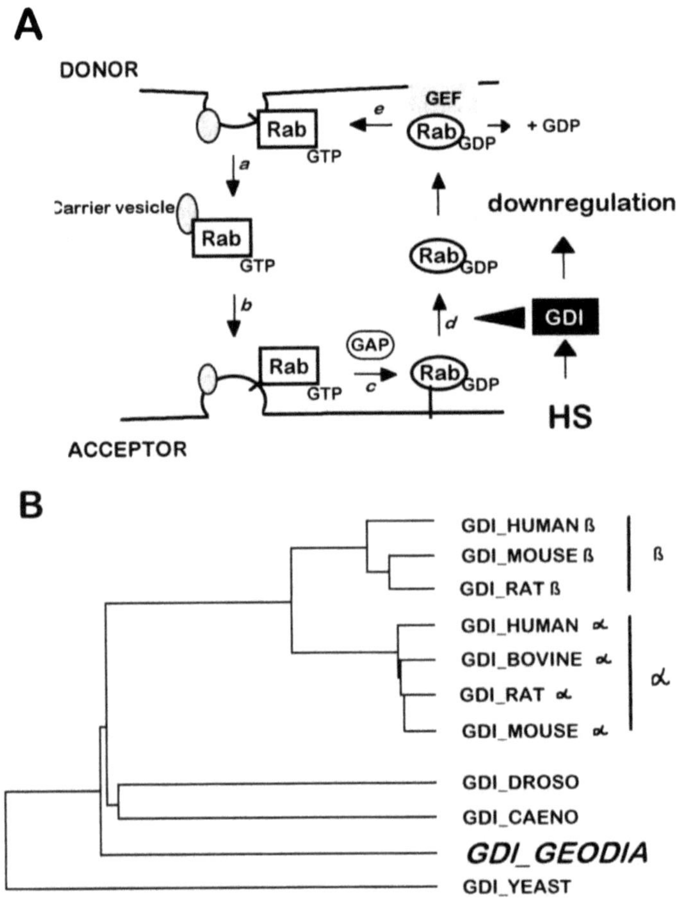

Fig. 10A–B. Rab/GDP dissociation inhibitor. A Scheme showing control of intracellular traffic by Rab protein. Rab protein in its active form is bound to GTP (*a*); it is anchored to the intracellular membrane via the geranylgeranyl group (donor site) and allows translocation of vesicles to the acceptor site (*b*). The GTP-bound Rab (active form) is slowly converted to the GDP-bound form (inactive), which is accelerated by the GTPase-activating protein(s) or GAPs (*c*). Subsequently, a cytosolic protein, the GDP dissociation inhibitor or GDI (*d*), removes GDP-associated Rab from the acceptor membrane and facilitates the transport of Rab to the donor membrane. Another component, the guanine nucleotide-exchange factor (*GEF*) catalyzes the exchange of GDP for GTP (*e*). When GDI is absent or inhibited, the intracellular traffic is halted. Upon heat stress, the expression of GDI is downregulated. **B** Phylogenetic tree based on a similarity-matrix analysis of the aa sequences of GDIs from the following organisms: sponge *G. cydonium* with human GDI β, mouse GDI β, rat GDI β, human GDI α, bovine GDI α, rat GDI α, mouse GDI α, *D. melanogaster* GDI (*GDI_DROSO*), *C. elegans* GDI (*GDI_CAENO*), and *S. cerevisiae* GDI (*GDI_YEAST*). The relationships, based on the number of similar aa between the species, are given

elegans GDI, *D, melanogaster* GDI, and *Saccharomyces cerevisiae* GDI (Fig. 10B). This comparison reveals that the sponge sequence is highly homologous to other GDIs, for example, the *G. cydonium* sequence shares with the human GDI β 285 identical and 130 similar aa out of a total number of 448 (*G.*

cydonium) / 445 (human) aa. In addition, it is shown that the *G. cydonium* molecule is the ancestor for the metazoan molecules.

5.9 Molecules Potentially Involved in Sponge Immunity

The following directions have been taken to elucidate molecules potentially involved in sponge immunity.

5.9.1 *Polymorphism in the Ig-Like Domains of the Receptor Tyrosine Kinase*

It was – parallel to higher vertebrates – intriguing to determine if the two immunoglobulin-like (Ig-like) domains of the RTK from *G. cydonium* display sequence polymorphism (Pancer et al. 1996a). The study design was as follows. Allo- and autografting experiments were performed using the two methods, parabiotic attachment and insertion technique. Thirty-six pairs of auto- and allografts were assayed. All of the autografts fused, whereas only two allografts fused and 34 pairs were incompatibile. Rejection among the parabiotic allografts was characterized by the formation of a collagenous barrier, whereas the allografts that were inserted into the host underwent destruction.

At the molecular level we analyzed RT-PCR products of amplification across the two Ig-like domains of RTK (about 500 bp) from two pairs of fusing sponges and one pair of rejecting sponges (Pancer et al. 1996a). Polymorphism was recorded, including 18 nt-substitution positions and a trinucleotide deletion, which translate into 13 polymorphic aa positions. Two of the six sponges were scored as heterozygotes. Among 9 informative polymorphic sites that were tested for linkage disequilibrium, 11 pairwise comparisons were found to be significant, implying the possibility of distinguishable alleles in this locus.

The aa alignment of the amplified region is shown in Fig. 11 and includes sequences from two fusible pairs, F-I (animals nos. 1 and 2; the sequences are abbreviated RTK1 and RTK2) and F-II (nos. 3 and 4; RTK3 and RTK4) as well as from the rejecting pair R-I (nos. 5 and 6; RTK5 and RTK6). In sponges nos. 1, 2, 3 and 6 the clones that were analyzed from each sample were identical, whereas in sponges nos. 4 and 5 two different sequences were found, named RTK4.1, RTK4.2, RTK5.1, and RTK5.2, respectively. Assuming that the genome of *G. cydonium* contains only a single locus coding for RTK, then animals nos. 4 and 5 are heterozygous at the RTK locus. Substitutions in 9 of the 13 polymorphic aa positions resulted in residues of similar physicochemical properties; the others displayed no alteration in the predicted secondary structure of the Ig-like domains. Hence it is reasonable to assume that the overall structure of the extracellular polymorphic domains remains unchanged.

The fact that only six individual sponges were sampled in this study, two pairs of fusions (nos. 1 and 2; nos. 3 and 4) and one pair of rejection (nos. 5 and

```
                                       Ig-like domain 1
                  ================================================================================
RTK1    F-I    1  PHSVTLTCSAASPPARDYQYQWQWRRNGTLLSNTHTRFSITPSTNTQSSSLVISGLRYSDAGDYMCTVEYGVCPDGVDCS
RTK2    F-I    1  ................................................................................
RTK3    F-II   1  ...............G................................................................
RTK4.1  F-II   1  ...............A.K......M..............................................A........
RTK4.2  F-II   1  ...............A.K......M.......................................................
RTK5.1  R-I    1  ...............G.......................................................A........
RTK5.2  R-I    1  ...................................-........S....N.....................A........
RTK6    R-I    1  ...............G.......................................................A........
                                 + +                          +    +                     +

                              Ig-like domain 1                    Ig-like domain 2
                  ============================================***********************************
RTK1    F-I   82  GTTPVTGVIHLDLPLIVEVDSSGLVVREGSEVVVLTCEVYGYPRDSSPPMWSSPGRNLESGRFNTTPRYTGTLSNGSVS
RTK2    F-I   82  ...............................................................................
RTK3    F-II  82  ................................I...............................I..............
RTK4.1  F-II  82  ................................I...............................I..............
RTK4.2  F-II  82  ................................I...............................I..............
RTK5.1  R-I   82  ................................I..............................................
RTK5.2  R-I   82  ................................I...............................I..............
RTK6    R-I   82  ......................E.....E.K.I.....................................S........
                                                + +                               +     +
```

Fig. 11. Polymorphism in the Ig-like domain of *G. cydonium* RTK. Multiple alignment of RT-PCR products of amplification across the Ig-like domains in TRK from six individuals. Sequences (five independent clones each have been analyzed) from fusing pairs *F-I* (animals nos. 1 and 2; the sequences are abbreviated *RTK1* and *RTK2*) and *F-II* (nos. 3 and 4; *RTK3* and *RTK4*), and from the rejecting pair *R-I* (animals no. 5 and 6; *RTK5* and *RTK6*) are compared. Two independent clones were sequenced from each individual; that are shown here only when different (*RTK4.1*, *RTK4.2*, *RTK5.1*, and *RTK5.2*). Only the deduced protein sequences are shown. Substitututions with respect to sequence RTK1 are shown; identities are marked (•). Substitutions of residues with similar physicochemical properties are indicated (+). The regions spanning Ig-like domain 1 (=) and 2 (★) are shown

6), precludes us from conclusions concerning possible linkage between the phenomenon of fusion/rejection in *G. cydonium* and the allelic polymorphism recorded in the Ig-like domains of the GCTK_2 locus. However, the data strongly indicate distinct polymorphic alleles in this locus.

To the best of our knowledge this is the first report of polymorphism in Ig-like domains of a receptor from invertebrates that may be associated with allorecognition. Further studies are needed to attest that fusion in sponges is confined to genetically identical individuals.

Parallel to this, sequencing studies were performed to determine if the second cell-attachment molecule hitherto identified in *G. cydonium*, the galectin, displays a similar polymorphism like Ig-like domains of RTK. Using the same animals, no aa change could be recorded.

5.9.2
Proteins and Genes in G. cydonium Related to the Human MHC

The human major histocompatibility complex (MHC) contains in addition to complement proteins also genes for the heat-shock protein HSP70, the cytosolic complex (proteasome), and the tumor necrosis factor (TNF); (Walter et al. 1994). The cDNAs coding for HSP70 and one proteasome subunit as well as the TNF activity, have been identified in *G. cydonium* (Fig. 12).

TNF. The activity of a TNF-like molecule was identified earlier (Pfeifer et al. 1992), suggesting that it plays a role in self/nonself recognition in *G. cydonium.*

HSP70. The cDNA coding for this heat-shock protein is known and was described above (Koziol et al. 1996).

Proteasome. The proteasome is a multicatalytic proteolytic system that appears to be an essential mechanism of cellular regulation, through degradation and turnover of ubiquitinated proteins. The proteasome is an unusually high molecular complex (~700 kDa) located in both the cytoplasm and the nucleus

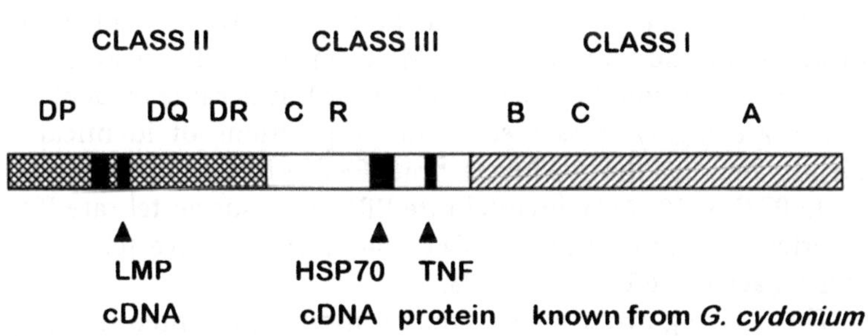

Fig. 12. The genes for heat-shock protein HSP70, proteasome subunits (*LMP*), and tumor necrosis factor (*TNF*) – their existence has been shown in *G. cydonium* – are present in humans inside the major histocompatibility complex

of all eukaryotic cells investigated so far, including yeast, plants, and animals (reviewed in Orlowski 1990; Peters 1994; Hilt and Wolf 1996) as well as in some species of bacteria and in archaebacteria (Seemüller et al. 1995). While the eukaryotic proteasome complex consists of a large particle of 26S (~1700 kDa) with a multicatalytic core protease of 20S, the bacterial proteasome consists only of the 20S particle that is formed by two sets of seven of α- and β-subunits. The 14 different eukaryotic proteasome subunits, which are all highly conserved, have been subgrouped by similarity to the bacterial subunits into seven α-type and seven β-type. It is currently accepted that the eukaryotic 20S proteasome consists of two homodimeric subcomplexes, each containing seven different members of the α- and β-subunits, with the exception of vertebrate β-subunits PRC7 and PRCC (also known as LMP2 and LMP7, respectively) which can replace homologous constitutive subunits of the 20S proteasome (reviewed in Hilt and Wolf 1996).

Most vertebrate cells present short peptides on MHC class I cell surface receptors, which are surveyed by cells of the immune system, primarily the T-cells. Substantial data now suggest that these antigenic peptides are the products of the proteasome-dependent machinery of degradation of intracellular proteins (Hilt and Wolf 1996). LMP2 and LMP7 (or PRC7 and PRCC), whose genes map in the MHC class II region (reviewed in Howard 1995), are interferon-γ inducible proteasome β subunits that upon induction can replace the homologous housekeeping subunits PRCD and PRCE (also known as Y and X) which are down-regulated (Akiyama et al. 1994). It appears that assimilation of the nonessential interferon-γ inducible subunits PRC7 and PRCC into the 20S proteasome contributes to increased hydrolysis after hydrophobic and basic aa residues, which alters the profile of the generated peptides into those favored for binding to MHC class I receptors (Howard 1995; Hilt and Wolf 1996).

We identified for the first time the sequence of two invertebrate proteasome subunits epsilon (PRCE) from the sponge *G. cydonium* and from the tunicate *Botryllus schlosseri*, whereas no evidence was found for the presence of a PRCC homologue (Pancer et al. 1996b).

Multiple alignment of two prokaryotic PRCBs (β-subunits of the 20 S complex) seven eukaryotic PRCEs including the two new invertebrate putative proteins and the four vertebrate MHC-encoded PRCCs available in the database, was performed (Pancer et al. 1996b). It became evident that these molecules are highly conserved, with 27 positions of identical residues. The identity between the two invertebrate PRCEs is 82%; invertebrate versus vertebrate PRCEs, 70–75%; invertebrate PRCEs versus vertebrate PRCCs, 68–70%; invertebrate versus yeast PRCEs, 64–71%; and invertebrate PRCEs versus archaebacterial PRCBs, 28–30%.

This striking evolutionary conservation is presented in a phylogram (Fig. 13), where the invertebrate sequences group together with the vertebrate PRCEs, the vertebrate PRCCs group in a different branch, and yeast PRCEs group together with the more distantly related prokaryotic PRCBs in a third

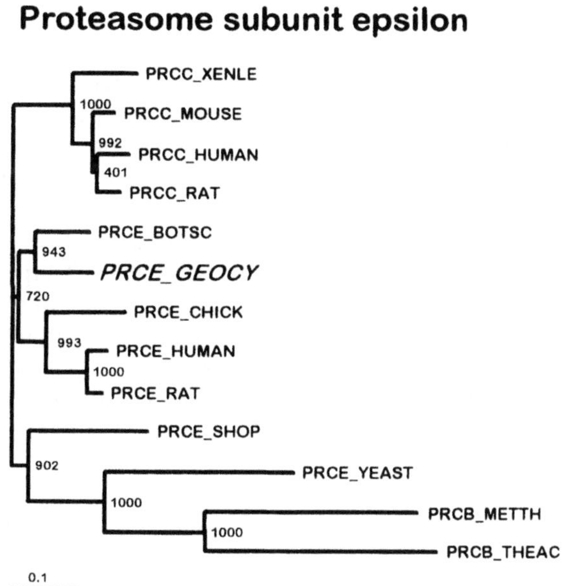

Fig. 13. Phylogram of the genetic distance between eukaryotic PRCEs: *PRCE_HUMAN, PRCE_RAT, PRCE_CHICK, PRCE_YEAST* (baker's yeast *S. cerevisiae*), *PRCE_SCHPO* (fission yeast *Schizosaccharomyces pombe*) with the two invertebrate sequences *PRCE_GEOCY* (*G. cydonium*), *PRCE_BOTSC* (tunicate *B. schlosseri*), the vertebrate MHC-encoded homologues PRCCs, *PRCC_RAT, PRCC_MOUSE, PRCC_HUMAN*, and *PRCC_XENLE* (*Xenopus laevis*), and prokaryotic PRCBs, *PRCB_METTH* (Archaebacterium *Methanosarcina thermophila*) and *PRCB_THEAC* (*Termoplasma acidophilum*). The phylogenetic tree was calculated with ClustalW. Bootstrap values at internal branching points are indicated (number of bootstrap trials 1000). *Scale bar* represents 10% distance index

branch. We therefore conclude that both of the invertebrate sequences are PRCE homologues, while no evidence was found for the presence of a PRCC homologue.

5.9.3
Proteins Featuring Scavenger Receptor Cysteine-Rich Domains

Proteins featuring scavenger receptor cysteine-rich (SRCR) domains comprise a superfamily, which includes one invertebrate and several vertebrate protein members (reviewed in Resnick et al. 1994). The SRCR domain consists of a 110 aa-residue motif with conserved spacing of six to eight cysteines, which apparently participate in intradomain disulfide bonds. Proteins of this superfamily feature 1 to 11 SRCR domain repeats (Resnick et al. 1994). The first molecule identified as a member of this superfamily was the sea urchin (*Strongylocentrotus purpuratus*) speract (sperm egg-peptide receptor) that featured four SRCR repeats (Dangott et al. 1989). Other members of the superfamily are so far only vertebrate sequences, such as macrophage type I scavenger receptor (Freeman et al. 1990), macrophage M130 antigen (Law et al. 1993), complement factor I (CFI; Goldberger et al. 1987), CD5 (Yu et al. 1990), CD6 (Aruffo et al. 1991), T-cell antigen WC1 (Wijngaard et al. 1992), cyclophilin-C-binding protein (Friedman et al. 1993) and its homologue MAC2-binding protein (Koths et al. 1993), a recently cloned sea lamprey (*Pteromyzon marinus*) protein (Mayer and Tichy 1995), a macrophage bacteria-binding receptor (MARCO; Elomaa et al. 1995), and CRP-ductin-α (Cheng et al. 1996). All cloned members of the SRCR superfamily are either cell-surface or secreted proteins, and most of them, with the exception of speract, appear to be involved in host-defense functions either expressed on T-cells, B-cells, and macrophages, or as soluble proteins (Resnick et al. 1994).

We identified the first putative SRCR protein from the marine sponge *G. cydonium* (Pancer et al. 1997b; Fig. 14). Three forms of SRCR molecules were

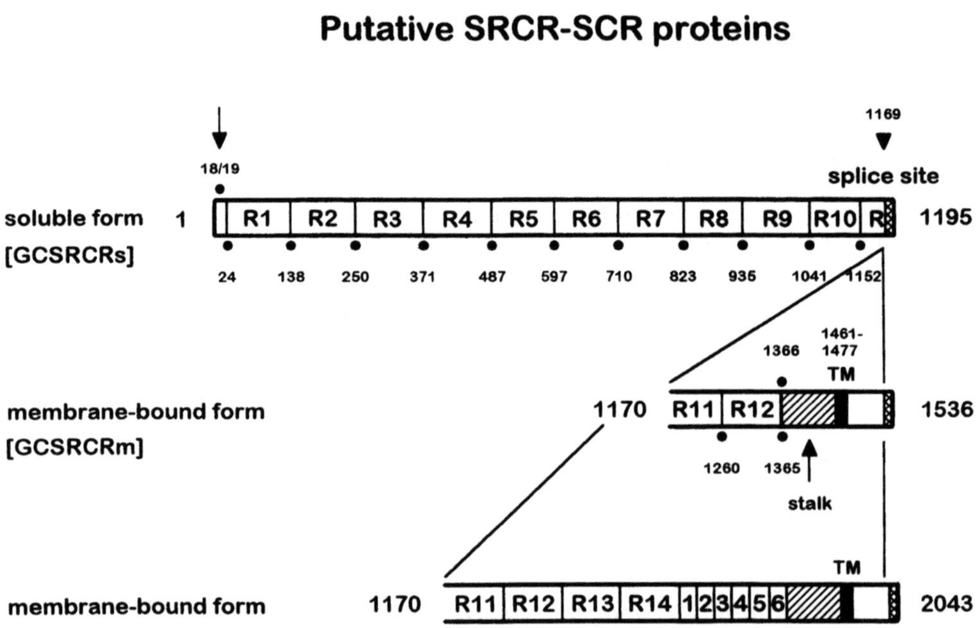

Fig. 14. Schematic presentation of the structural features in *G. cydonium* putative *SRCR-SCR* proteins. Three forms of GCSRCR protein were characterized: *GCSRCRs* (the soluble form), which features ten SRCR repeats; *GCSRCRm* (the membrane-bound form; only its C-terminus is shown) with 12 SRCR repeats, a stalk-like Ser and Pro-rich segment (*hatched*), a C-terminal transmembrane domain (*TM; filled box*), and a cytoplasmic tail; as well as the *GCSRCR-SCRm* molecule which contains in addition six SCR segments. GCSRCRs lacks the two C-terminal SRCR repeats and the transmembrane domain, but both forms share the last 26 residues (*cross-hatched*) in common. Presumably, these forms represent events of alternative splicing of the same transcript. The GCSRCR-SCRm molecule comprises the cell attachment signature RGD

characterized, which apparently represent alternative splicing of the same transcript. The long putative SRCR protein, 1536 aa, features 12 SRCR repeats, a C-terminal transmembrane domain, and a cytoplasmic tail. The sequence of the short form is identical to the long form, except that it lacks a coding region near the C-terminus; thus the 1195 aa-deduced protein consists of only the first 10 SRCR domains and the last 26 C-terminal aa residues without the transmembrane domain. Homology searches revealed that the sponge putative SRCR protein shares with bovine T-cell antigen WC1 29.2% identity in 1054 aa overlap, 33.9% identity in 475 aa overlap with sea urchin speract, and 56% identity in 110 aa overlap with macrophage scavenger receptor type I. Based on the number and location of the conserved Cys residues, the sponge SRCR repeats were classified as belonging to group A of the SRCR superfamily. With 12 SRCR repeats, one more than those in any of the previously described SRCR proteins, and with several membrane-bound and soluble forms, it seems that the most primitive known member of this family may be among the most structurally compound ones.

5.9.4 *Molecules Comprising Short Consensus Repeats*

In the course of elucidating further splice forms of the sponge putative SRCR protein, another sequence was identified which contained, in addition to the 12 SRCR repeats, 2 further repeats which are linked to 6 short consensus repeats (SCR), the SRCR-SCR molecule (Fig. 14). Like SRCR proteins, SCR proteins are primarily found in the family of complement-regulatory proteins (Zipfel and Skerka 1994). Studies are now ongoing to solve the question of whether the "polymorphism" seen in the *G. cydonium* SRCR-SCR is typical for the species only or is linked to an immune function.

6 Evolutionary Tempo

In molecular evolution, one unsolved question is the validity of the molecular clock hypothesis, which implies that the rate of molecular evolution is nearly constant per year among different evolutionary lineages (Kimura 1983; Li et al. 1985, 1987). Consequently, this rate should be linked to the mutation rate and hence be more closely correlated with the number of generations per unit time than to time itself. However, the standard value for the average rate of nonsynonymous substitutions in DNA can vary between 0.6 and 0.9×10^{-9} per site and year (Li et al. 1985, 1987). Therefore, the calculated time of divergence of the different phyla might vary as well.

For the first time, aa sequences from a sponge (*G. cydonium*) have been used to calculate the time of divergence of Porifera from Metazoa, firstly, the conserved aa sequence in the tyrosine kinase domains of the RTK, and, secondly, the sequences of two sponge S-type lectins. It was estimated that the sponge lectin diverged from the common ancestral gene for S-type lectins 800 Ma ago (Precambrian era; Hirabayashi and Kasai 1993). A somewhat lower value was found for RTKs (see above). The figures derived from these sequences are in good agreement with the generally accepted time of divergence found by traditional methods; see Sections 5.3.3 and 5.4.1. Nevertheless, the estimated time of divergence of sponge proteins from a common ancestor of other metazoan organisms is valid only if we assume constancy of the rate of aa substitution during molecular evolution.

With respect to the evolutionary tempo, paleontological data revealed that, on the time scale of eukaryotic evolution, the tempo has increased episodically (Knoll 1994). For the earliest recorded period of protist evolution (>1700 Ma ago), the morphological diversity and the turnover rates were low. With the appearance of heterotrophs as well as of algae (Meso-/Neoproterozoic boundary; 1000 Ma ago) both the diversity and the turnover rates increased. A further acceleration was recorded with the appearance of radiating animals (Proterozoic/Early Cambrian; 500 Ma ago).

7 Conclusion

Until recently, sponges have been regarded as colonies of unspecialized cells – "individual flagellates" – which require only cells that secrete adhesive glycoprotein and bind to it (Loomis 1988). This view has completely changed with the application of modern molecular biological techniques. The results – accumulated just during the past 3 years and reviewed here – reveal that sponges are composed of molecules which are typical for Metazoa. It appears that all basic molecules required for the maintenance of an individuum as known from higher metazoan phyla are already present in sponges. Now, three questions arise: first, how can we describe the evolutionary trend from the relatively simple phylum porifera to the highly developed Chordata or Arthropoda in molecular term? Second, what is (are) the driving force(s) of this evolutionary process? And third, which underlying mechanisms may lead to a directional evolution to a given phylum?

A further question as to the general phylogeny of Metazoa remains unsolved, the question regarding the evolution of the adhesion and receptor molecules in the protozoan lineage. It is unlikely that these molecules evolved suddenly. Focusing on the sponge receptor tyrosine kinase, the structure of which consists of several domains, indicating that this receptor is already highly complex and very likely composed of different building blocks formed during evolution. For the development of these elements in the eukaryotic protist lineage, a period of ≈1000 Ma – according to present estimates – was necessary. An experimental approach to answer this question is given in a further chapter of this book (Müller and Müller, this vol.).

Acknowledgements. I am grateful to my coworkers for their contributions, especially C. Koziol, A. Krasko, M. Kruse, Z. Pancer, A. Skorokhod, and C. Wagner. Support by the Deutsche Forschungsgemeinschaft (Mü 348/13-1), the Stiftung Volkswagenwerk, the Fonds der Chemischen Industrie (0161191) and the International Human Frontier Science Program (RG-333/96-M) is gratefully acknowledged.

References

Agnès F, Shamoon B, Dina C, Rosnet O, Birnbaum D, Galibert F (1994) Genomic structure of the downstream part of the human FLT3 gene: exon/intron conservation among genes encoding receptor tyrosine kinase (RTK) of subclass III. Gene 145: 283–288

Akiyama K, Yokota K, Kagawa S, Shimbara N, Tamura T, Akioka H, Nothwang HG, Noda C, Tanaka K, Ichihara A (1994) cDNA cloning and interferon γ down-regulation of proteasomal subunits X and Y. Science 265: 1231–1234

Allen BG, Walsh WP (1994) The biochemical basis of the regulation of smooth-muscle contraction. Trends Biochem Sci 19: 362–368

Aruffo A, Melnick MB, Linsley PS, Seed B (1991) The lymphocyte glycoprotein CD6 contains a repeated domain structure characteristic of a new family of cell surface and secreted proteins. J Exp Med 174: 949–952

Barnes RD (1980) Invertebrate zoology, 4th edn. Saunders, Philadelphia

Biesalski HK, Doepner G, Tzimas G, Gamulin V, Schröder HC, Batel R, Nau H, Müller WEG (1992) Modulation of *myb* gene expression in sponges by retinoic acid. Oncogene 7: 1765–1774

Boguski MS, McCormick F (1993) Proteins regulating Ras and its relatives. Nature 366: 643–654

Bork P, Sander C, Valencia A, Bukau B (1992) A module of the DnaJ heat shock proteins found in malaria parasites. Trends Biochem Sci 17: 129

Britten R, Davidson EH (1969) Gene regulation for higher cells: a theory. Science 165: 349–357

Brundage RA, Fogarty KE, Tuft RA, Fay FS (1991) Calcium gradients underlying polarization and chemotaxis of eosinophiles. Science 254: 703–706

Buscema M, Van de Vyver G (1983) Variability of allograft rejection processes in *Axinella verrucosa*. Dev Comp Immunol 7: 613–616

Campbell NA (1996) Biology. Benjamin Cummings, Menlo Park

Cavalier-Smith T (1985) Eukaryotic gene number, non-coding DNA and genome size. In: Cavalier-Smith T (ed) The evolution of genome size. Wiley, New York

Cavalier-Smith T (1991) Intron phylogeny: a new hypothesis. Trends Genet 7: 145–148

Cavalier-Smith T, Allsopp MTEP, Chao EE, Boury-Esnault N, Vacelet (1996) Sponge phylogeny, animal monophyly, and the origin of the nervous system: 18S rRNA evidence. Can J Zool 74: 2031–2045

Cheng H, Bjerknes M, Chen H (1996) CRP-ductin: gene expressed in intestinal crypts and in pancreatic and hepatic ducts. Anat Rec 244: 327–343

Christen R, Ratto A, Baroin A, Perasso R, Grell KG, Adoutte A (1991) An analysis of the origin of metazoans, using comparisons of partial sequences of the 28S RNA, reveals an early emergence of triploblasts. EMBO J 10: 499–503

Coutinho CC, Vissers S, Van de Vyver G (1994) Evidence of homeobox genes in the freshwater sponge *Ephydatia fluviatilis*. In: Soest RWM van, Kempen TMG van, Braekman JC (eds) Sponges in time and space. Balkema Press, Rotterdam, pp 385–388

Dangott LJ, Jordan JE, Bellet RA, Galbers DL (1989) Cloning of the mRNA for the protein that crosslinks to the egg peptide speract. Proc Natl Acad Sci USA 86: 2128–2132

Darnell JE (1978) Implications of RNA-RNA splicing in evolution of eukaryotic cells. Science 202: 1257–1260

De Robertis EM, Morita EA, Cho KWY (1991) Gradient fields and homeobox genes. Development 112: 669–678

Delage Y (1892) Embryogenèse des éponges siliceuses. Arch Zool Exp Gén 10: 345–498
Diehl-Seifert B, Uhlenbruck G, Geisert M, Zahn RK, Müller WEG (1985) Physicochemical and functional characterization of the polymerization process of the *Geodia cydonium* lectin. Eur J Biochem 147: 517–523
Doolittle WF (1978) Genes in pieces: were they ever together? Nature 272: 581–582
Doolittle WF, Brown JR (1994) Tempo, mode, the progenote, and the universal root. Proc Natl Acad Sci USA 91: 6735–6742
Elomaa O, Kangas M, Sahlberg C, Tuukkanen J, Sormunen R, Liakka A, Thesleff I, Kraal G, Tryggvason K (1995) Cloning of a novel bacteria-binding receptor structurally related to scavenger receptors and expressed in a subset of macrophages. Cell 80: 603–609
Exposito JY, Garrone R (1990) Characterization of a fibrillar collagen gene in sponges reveals the early evolutionary appearance of two collagen families. Proc Natl Acad Sci USA 87: 6669–6673
Field KG, Olsen GJ, Lane DJ, Giovannoni SJ, Ghiselin MT, Raff EC, Pace NR, Raff RA (1988) Molecular phylogeny of the animal kingdom. Science 239: 748–753
Freeman M, Ashkenas J, Rees DJ, Kingsley DM, Copeland NG, Jenkins NA, Krieger M (1990) An ancient, highly conserved family of cysteine-rich protein domains revealed by cloning type I and type II murine macrophage scavenger receptors. Proc Natl Acad Sci USA 87: 8810–8814
Friedman J, Trahey M, Weissman I (1993) Cloning and characterization of cyclophilin C-associated protein: a candidate for natural cellular ligand for cyclophilin C. Proc Natl Acad Sci USA 90: 6815–6819
Gamulin V, Rinkevich B, Schäcke H, Kruse M, Müller IM, Müller WEG (1994) Cell adhesion receptors and nuclear receptors are highly conserved from the lowest Metazoa [marine sponges] to vertebrates. Biol Chem Hoppe-Seyler 375: 583–588
Gamulin V, Skorokhod A, Kavsan V, Müller IM, Müller WEG (1997) Experimental indication against blockwise evolution of metazoan protein molecules: example, receptor tyrosine kinase gene from the sponge *Geodia cydonium*. J Mol Evol 44: 242–252
Geer P, Hunter T, Lindberg RA (1994) Receptor protein-tyrosine kinases and their signal transduction pathways. Annu Rev Cell Biol 10: 251–337
Georgopoulos C, Liberek K, Zylicz M, Ang D (1994) Properties of the heat shock proteins of *Escherichia coli* and the autoregulation of the heat shock response. In: Morimoto RI, Tissères A, Georgopoulos C (eds) The biology of heat shock proteins and molecular chaperones. Cold Spring Harbor Laboratory Press, Cold Spring Harbor, New York, pp 209–249
Gilbert W (1987) The exon theory of genes. Cold Spring Harbor Symp Quant Biol 52: 901–905
Gilbert W, Marchionni M, McKnight G (1986) On the antiquity of introns. Cell 46: 151–154
Goldberger G, Bruns GA, Rits M, Edge MD, Kwiatkowski DJ (1987) Human complement factor I: analysis of cDNA-derived primary structure and assignment of its gene to chromosome 4. J Biol Chem 262: 10065–10071
Gramzow M, Schröder HC, Fritsche U, Kurelec B, Robitzki A, Zimmermann H, Friese K, Kreuter MH, Müller WEG (1989) Role of phospholipase A2 in the stimulation of sponge cell proliferation by the homologous lectin. Cell 59: 939–948
Greenberg ME, Ziff EB (1984) Stimulation of 3T3 cells induces transcription of the c-*fos* proto-oncogene. Nature 311: 433–438
Grotzinger JP, Bowring SA, Saylor B, Kauffmann AJ (1995) New biostratigraphic and geochronologic constraints on early animal evolution. Science 270: 598–604
Hadzi J (1963) The evolution of the Metazoa. Pergamon Press, Oxford
Haeckel E (1866) Generelle Morphologie der Organismen, Zweiter Band. Allgemeine Entwicklungsgeschichte der Organismen. Georg Reimer, Berlin
Haeckel E (1868) Natürliche Schöpfungsgeschichte. Georg Reimer, Berlin, pp 383–432
Haga T, Haga K, Hulme CE (1990) Solubilization, purification, and molecular characterization of receptors: principles and strategy. In: Hulme EC (ed) Receptor biochemistry. IRL Press, Oxford, pp 1–50
Halanych KM (1991) 5S ribosomal RNA sequences inappropriate for phylogenetic reconstruction. Mol Biol Evol 8: 249–253

Hardie G, Hanks S (1995) The protein kinase FactsBook: protein-tyrosine kinases. Academic Press, London
Henkart P, Humphreys S, Humphreys T (1973) Characterization of sponge aggregation factor. A unique proteoglycan complex. Biochemistry 12: 3045–3050
Hightower LE, Sadis SE, Takenaka, IM (1994) Interactions of vertebrate hsc70 and hsp70 with unfolded proteins and peptides. In: Morimoto RI, Tissières A, Georgopoulus C (eds) The biology of heat shock proteins and molecular chaperones. Cold Spring Harbor Laboratory Press, Cold Spring Harbor Laboratory, New York pp 179–207
Hildemann WH, Johnston IS, Jokiel PL (1979) Immunocompetence in the lowest metazoan phylum: transplantation immunity in sponges. Science 204: 420–422
Hilt W, Wolf DH (1996) Proteasomes: destruction as a programme. Trends Biochem Sci 21: 96–102
Hirabayashi J, Kasai K (1993) The family of metazoan metal-independent β-galactoside-binding lectins: structure, function and molecular evolution. Glycobiology 3: 297–304
Howard JC (1995) Supply and transport of peptides presented by class I MHC molecules. Curr Biol 7: 69–76
Hunkapiller T, Goverman J, Koop BF, Hood L (1989) Implications of the diversity of the immunoglobulin gene superfamily. Cold Spring Harbor Symp Quant Biol 54: 15–29
Hunter T, Lindberg RA, Middlemas DS, Tracy S, Geer P (1992) Receptor protein kinases and phosphatases. Cold Spring Harbor Symp Quant Biol 58: 25–41
Hyman LH (1951) The invertebrates. McGraw-Hill, New York
Imsiecke G, Borojevic R, Custodio M, Müller WEG (1994) Retinoic acid acts as a morphogen in freshwater sponges. Invertebr Reprod Dev 26: 89–98
Imsiecke G, Custodio M, Borojevic R, Steffen R, Moustafa MA, Müller WEG (1995) Genome size and chromosomes in marine sponges [*Suberites domuncula*, *Geodia cydonium*]. Cell Biol Intern 19: 995–1000
Iwabe N, Kuma K, Hasegawa M, Osawa S, Miyata T (1989) Evolutionary relationship of archaebacteria, eubacteria, and eukaryotes inferred from phylogenetic trees and duplicated genes. Proc Natl Acad Sci USA 86: 9355–9359
Jaenicke R (1994) Eye-lens proteins: structure, superstructure, stability, genetics. Naturwissenschaft 81: 423–429
Jameson R (1811) Catalogue of animals of the class of Vermes, found in the Firth of Forth and other parts of Scotland. Mem Werner Soc 1: 556–565
Kasai KI, Hirabayashi J (1996) Galectins: a family of animal lectins that decipher glycocodes. J Biochem 119: 1–8
Kent S (1878) Notes on the embryology of sponges. Ann Mag Nat Hist 5, 2: 139–156
Kimura M (1983) The neutral theory of molecular evolution. Cambridge University Press, Cambridge
Knoll AH (1994) Proterozoic and Early Cambrian protists: evidence for accelerating evolutionary tempo. Proc Natl Acad Sci USA 91: 6743–6750
Kodama R, Eguchi G (1994) Gene regulation and differentiation in vertebrate ocular tissues. Curr Opin Genet Dev 4: 703–708
Koths K, Taylor E, Halenbeck R, Casipit C, Wang A (1993) Cloning and characterization of a human Mac-2-binding protein, a new member of the superfamily defined by the macrophage scavenger receptor cysteine-rich domain. J Biol Chem 268: 14225–14249
Koziol C, Wagner-Hülsmann C, Mikoc A, Gamulin V, Kruse M, Pancer Z, Schäcke H, Müller WEG (1996) Cloning of a heat-inducible biomarker, the cDNA encoding the 70-kDa heat shock protein, from the marine sponge *Geodia cydonium*: response to natural stressors. Mar Ecol Prog Ser 136: 153–161
Koziol C, Kruse M, Batel R, Badria FA, Müller IM, Müller WEG (1997) Characterization of cDNAs encoding DnaJ from the marine sponge *Geodia cydonium*: increased gene expression in response to heat shock. Mar Ecol Progr Ser; in press
Krasko A, Müller WEG (1997) Evolutionary relationships of the metazoan βγ-crystallins, including the one from the marine sponge *Geodia cydonium*. Proc R Soc Biol Sci (in press)

Krasko A, Scheffer U, Koziol C, Pancer Z, Batel R, Badria FA, Müller WEG (1997) Diagnosis of sublethal stress in the marine sponge *Geodia cydonium*: application of the 70-kDa heat shock protein and of the novel biomarker, the Rab GDP-dissociation inhibitor, as a probe. Aquat Toxicol 37: 157–168

Kruse M, Gamulin V, Cetkovic H, Pancer Z, Müeller IM, Müller WEG (1996) Molecular evolution of the metazoan protein kinase C multigene family. J Mol Evol 43: 374–383

Kuhns WJ, Popescu O, Burger MM, Misevic G (1995) Sulfate restriction induces hypersecretion of the adhesion proteoglycan and cell hypomotility associated with increased $^{35}SO_4^{2-}$ uptake and expression of a band 3 like protein in the marine sponge *Microciona prolifera*. J Cell Biochem 57: 71–89

Labat-Robert J, Robert L, Auger C, Lethias C, Garrone R (1981) Fibronectin-like protein in Porifera: its role in cell aggregation. Proc Natl Acad Sci USA 78: 6261–6265

Lafay B, Boury-Esnault N, Vacelet J, Christen R (1992) An analysis of partial 28S ribosomal RNA sequences suggests early radiation of sponges. BioSystems 28: 139–151

Law SK, Micklem KJ, Shaw JM, Zhang XP, Dong Y, Willis AC, Mason DY (1993) A new macrophage differentiation antigen which is a member of the scavenger receptor superfamily. Eur J Immunol 23: 2320–2325

Ledger PW (1975) Septate junctions in the calcareous sponge *Syncon ciliatum*. Tissue Cell 7: 13–18

Lendenfeld R v (1903) Tetraxonia. In: Das Tierreich - Porifera, 19. Lief Reimer, Berlin, pp 113–227

Lentz TL (1966) Histochemical localization of neurohumors in a sponge. J Exp Zool 162: 171–180

Li WH, Graur D (1991) Fundamentals in molecular evolution. Sinnauer Assoc, Sunderland, Massachusetts

Li WH, Luo CC, Wu CI (1985) Evolution of DNA sequences. In: MacIntyre RJ (ed) Molecular evolutionary genetics. Plenum Press, New York, pp 1–94

Li WH, Tanimura M, Sharp PM (1987) An evaluation of the molecular clock hypothesis using mammalian DNA sequences. J Mol Evol 25: 330–342

Loomis WF (1988) Four billion years: an assay on the evolution of genes and organisms. Sinnauer Assoc, Sunderland, Massachusetts

Lorenz B, Bohnensack R, Gamulin V, Steffen R, Müller WEG (1996) Regulation of motility of cells from marine sponges by calcium ions. Cell Signal 8: 517–524

Martin GR, Timpl R (1987) Laminin and other basement membrane components. Annu Rev Cell Biol 3: 57–85

Mayer RJ, Arnold J, Laszlo L, Landon M, Lowe J (1991) Ubiquitin in health and disease. Biochim Biophys Acta 1089: 141–157

Mayer WE, Tichy H (1995) A cDNA clone from the sea lamprey *Pteromyzon marinus* coding for a scavenger receptor Cys-rich (SRCR) domain protein. Gene 164: 267–271

McLachlan AD, Karn J (1982) Periodic charge distributions in the myosin rod amino acid sequence match cross-bridge spacings in muscle. Nature 299: 226–231

Mehl D, Müller I, Müller WEG (1997) Molecular biological and palaeontological evidence that Eumetazoa, including Porifera (sponges), are of monophyletic origin. In: Proc Int Conf Sponge Science, 12–16 March 1996, Otsu, Japan. Springer, Berlin Aeidelberg New York (in press)

Mehlhorn H (ed) (1989) Grundriβ der Zoologie. Fischer, Stuttgart

Merck KB, Groenen PJTA, Voorter CEM, de Haard-Hoekman WA, Horwitz J, Bloemendal H, de Jong WW (1993) Structural and functional similarities of bovine α-crystallin and mouse small heat-shock protein: a family of chaperones. J Biol Chem 268: 1046–1052

Morris PL (1993) The developmental role of the extracellular matrix suggests a monophyletic origin of the kingdom Animalia. Evolution 47: 152–165

Mukai H (1992) Allogeneic recognition and sex differentiation in chimeras of the freshwater sponge *Ephydatia muelleri*. J Exp Zool 264: 298–311

Müller WEG (1982) Cell membranes in sponges. Int Rev Cytol 77: 129–181

Müller WEG (1995) Molecular phylogeny of Metazoa [animals]: monophyletic origin. Naturwissenschaft 82: 321–329
Müller WEG, Zahn RK (1973) Purification and characterization of a species-specific aggregation factor in sponges. Exp Cell Res 80: 95–104
Müller WEG, Schäcke H (1996) Characterization of the receptor protein-tyrosine kinase gene from the marine sponge *Geodia cydonium*. Prog Mol Subcell Biol 17: 183–208
Müller WEG, Müller IM (1997) Primordial endocrinology in the lowest invertebrates: the porifera. In: Dorn A (ed) Reproductive biology of invertebrates. J Wiley, Chichester (in press)
Müller WEG, Müller I, Zahn RK, Kurelec B (1976) Species-specific aggregation factor in sponges; VI. aggregation receptor from the cell surface. J Cell Sci 21: 227–241
Müller WEG, Rottmann M, Diehl-Seifer B, Kurelec B, Uhlenbruck G, Schröder HC (1987) Role of the aggregation factor in the regulation of phosphoinositide metabolism in sponges. Possible consequences on calcium efflux and on mitogenesis. J Biol Chem 262: 9850–9858
Müller WEG, Müller IM, Gamulin V (1994) On the monophyletic evolution of the Metazoa. Braz J Med Biol Res 27: 2083–2096
Müller WEG, Müller IM, Rinkevich B, Gamulin V (1995) Molecular evolution: evidence for the monophyletic origin of multicellular animals. Naturwissenschaft 82: 36–38
Müller WEG, Steffen R, Rinkevich B, Matranga V, Kurelec B (1996) Multixenobiotic resistance mechanism in the marine sponge *Suberites domuncula*: its potential applicability for the evaluation of environmental pollution. Mar Biol 125: 165–170
Müller WEG, Blumbach B, Wagner-Hülsmann C, Lessel U (1997) Galectins in the phylogenetically oldest metazoa, the sponges [Porifera]. Trends Glycosci Glycotechnol 9: 123–130
Nielsen C (1995) Animal evolution. Oxford University Press, Oxford
Orgel LE, Crick FH (1980) Selfish DNA: the ultimate parasite. Nature 284: 604–607
Orlowski M (1990) The multicatalytic proteinase complex, a major extralysosomal proteolytic system. Biochemistry 29: 10289–10297
Pancer Z, Kruse M, Schäcke H, Scheffer U, Steffen R, Kovács P, Müller WEG (1996a) Polymorphism in the immunoglobulin-like domains of the receptor tyrosine kinase from the sponge *Geodia cydonium*. Cell Adhes Commun 4: 327–339
Pancer Z, Scheffer U, Müller I, Müller WEG (1996b) Cloning of sponge (*Geodia cydonium*) and tunicate (*Botryllus schlosseri*) proteasome subunit epsilon (PRCE): implications about the evolution of the vertebrate MHC-encoded homologue LMP7 (PRCC). Biochem Biophys Res Commun 228: 406–410
Pancer Z, Müller WEG (submitted) Extreme sequence divergence in the intron that splits the two Ig-like domains of RTK, within and among individuals: identification of a hyperactive evolutionary site in the marine sponge *Geodia cydonium*
Pancer Z, Kruse M, Müller I, Müeller WEG (1997a) On the origin of adhesion receptors of metazoa: cloning of the integrin α subunit cDNA from the sponge *Geodia cydonium* Mol Biol Evol 14: 391–398
Pancer Z, Münkner J, Müller I, Müller WEG (1997b) A novel member of an ancient superfamily: sponge (*Geodia cydonium*, Porifera) putative protein that features scavenger receptor cysteine-rich repeats Gene (in press)
Peters JM (1994) Proteasomes: protein degradation machines of the cell. Trends Biochem Sci 19: 377–382
Pfeifer K, Schröder HC, Rinkevich B, Uhlenbruck G, Hanisch F-G, Kurelec B, Scholz P, Müller WEG (1992) Immunological and biological identification of tumor necrosis factor in sponges: role of this factor in the formation of necrosis in xenografts. Cytokine 4: 161–169
Pfeifer K, Haasemann M, Gamulin V, Bretting H, Fahrenholz F, Müller WEG (1993a) S-type lectins occur also in invertebrates: high conservation of the carbohydrate recognition domain in the lectin genes from the marine sponge *Geodia cydonium*. Glycobiology 3: 179–184

Pfeifer K, Frank W, Schröder HC, Gamulin V, Rinkevich B, Batel R, Müller IM, Müller WEG (1993b) Cloning of the polyubiquitin gene from the marine sponge *Geodia cydonium* which is preferentially expressed during reaggregation of cells. J Cell Sci 106: 545–554

Rasmont R (1970) Some new aspects of the physiology of fresh-water sponges. In: Fry WG (ed) The biology of the Porifera. Symp Zool Soc Lond 25: 415–422

Reiswig H (1971) In situ pumping activity of tropical Demospongiae. Mar Biol 9: 38–50

Reitner J, Mehl D (1995) Early Paleozoic diversification of sponges: new data and evidences. Geol Paläont Mitt Innsbruck 20: 335–347

Resnick D, Pearson A, Krieger M (1994) The SRCR superfamily: a family reminiscent of the Ig superfamily. Trends Biochem Sci 19: 5–8

Rodrigo AG, Bergquist PR, Bergquist PL, Reeves RA (1994) Are sponges animals? An investigation into the vagaries of phylogenetic interference. In: van Soest RWM, van Kempen TMG, Braekman JC (eds) Sponges in time and space. Balkema Press, Rotterdam, pp 47–54

Rousset D, Agnès F, Lachaume P, André C, Galibert F (1995) Molecular evolution of the genes encoding receptor tyrosine kinase with immunoglobulinlike domains. J Mol Evol 41: 421–429

Salvini-Plawen L (1978) On the origin and evolution of the lower Metazoa. Z Zool Sys Evolutionsforsch 16: 40–88

Sarà M, Vacelet J (1973) Ecologie des démosponges, Traité de zoologie: Spongiaires. Tome III (1) Grassé PP (ed), Masson, Paris, pp 462–576

Schäcke H, Rinkevich B, Gamulin V, Müller IM, Müller WEG (1994a) Immunoglobulin-like domain is present in the extracellular part of the receptor tyrosine kinase from the marine sponge *Geodia cydonium*. J Mol Recognit 7: 272–276

Schäcke H, Schröder HC, Gamulin V, Rinkevich B, Müller IM, Müller WEG (1994b) Molecular cloning of a tyrosine kinase gene from the marine sponge *Geodia cydonium*: a new member belonging to the receptor tyrosine kinase class II family. Mol Membr Biol 11: 101–107

Schäcke H, Schröder HC, Müller IM, Müller WEG, Gamulin V, Rinkevich B (1994c) The immunoglobulin superfamily includes members from the lowest invertebrates to the highest vertebrates. Immunol Today 15: 497–498

Schäcke H, Müller IM, Müller WEG (1994d) Tyrosine kinase from the marine sponge *Geodia cydonium*: the oldest member belonging to the receptor tyrosine kinase class II family. In: Müller WEG (ed) Use of aquatic invertebrates as tools for monitoring of environmental hazards. Gustav Fischer, Stuttgart, pp 200–211

Scheffer U, Krasko A, Pancer Z, Müller WEG (1997a) Isolation and characterization of the cDNA clone encoding the serum response factor homolog in the marine sponge *Geodia cydonium*: high conservation of this transcription factor within Metazoa. Biol J Linn Soc (in press)

Scheffer U, Koziol C, Pancer Z, Krasko A, Müller WEG (1997b) Sponges as biomarkers of the aquatic environment: application of molecular probes. In: Proc Int Conf Sponge Science, 12–16 March 1996, Otsu, Japan. Springer, Berlin Heidelberg New York (in press)

Schlesinger MJ, Ashburner M, Tissieres A (1982) Heat shock. From bacteria to man. Cold Spring Harbor Laboratory Press, Cold Spring Harbor, New York

Schopf JW (1994) The oldest known records of life: early Archean stromatolites, microfossils, and organic matter. In: Bengtson S (ed) Early life on earty. Columbia University Press, New York, pp 193–206

Schopf JW, Haugh BN, Molnar RE, Satterthwait DF (1973) On the development of metaphytes and metazoans. J Paleontol 47: 1–9

Schwartz MA, Schaller MD, Ginsberg MK (1995) Integrins: emerging paradigms of the signal transduction. Annu Rev Cell Dev Biol 11: 9–99

Seemüller E, Lupas A, Stock D, Löwe J, Huber R, Baumeister W (1995) Proteasome from *Thermoplasma acidophilum*: a threonine protease. Science 268: 579–582

Seimiya A, Ishiguro H, Miura K, Watanabe Y, Kurosawa Y (1994) Homeobox-containing genes in the most primitive metazoa, the sponges. Eur J Biochem 221: 219–225

Seino S, Seino M, Nishi S, Bell GI (1989) Structure of the human insulin receptor gene and characterization of its promoter. Proc Natl Acad Sci USA 86: 114–118

Shore P, Sharrocks AD (1995) The MADS-box family of transcription factors. Eur J Biochem 229: 1–13

Simon K, Zerial M (1993) Rab proteins and the road maps for intracellular transport. Neuron 11: 789–799

Simpson TL (1984) The cell biology of sponges. Springer, Berlin Heidelberg New York

Smith LC, Hildemann WH (1986) Allogeneic cell interactions during graft rejection in *Callyspongia diffusa* (Porifera; Demospongia); a study with monoclonal antibodies. Proc R Soc Lond B 226: 465–377

Smothers JF, Dohlen CD v, Smith LH, Spall RD (1994) Molecular evidence that the Myxozoan protists are metazoans. Science 265: 1719–1721

Sollas WJ (1884) On the development of *Halisarca lobularis* (O. Schmidt) Q J 24: 603–612

Steiner M, Mehl D, Reitner J, Erdtmann BD (1993) Oldest entirely preserved sponges and other fossils from the Lowermost Cambrian and a new facies reconstruction of the Yangtze Platform (China). Ber Geowiss Abh (E) 9: 293–329

Sun H, Tonks NK (1994) The coordinated action of protein tyrosine phosphatases and kinases in cell signaling. Trends Biochem Sci 19: 480–485

Szathmary E, Maynard Smith J (1995) The major evolutionary transitions. Nature 374: 227–232

Takaishi K, Sasaki T, Karneyama T, Tsukita A, Takai Y (1995) Translocation of activated *Rho* from the cytoplasm to membrane ruffling area, cell- adhesion sites and cleavage furrows. Oncogene 11: 39–48

Tan JL, Ravid S, Spudich JA (1992) Control of nonmuscle myosins by phosphorylation. Annu Rev Biochem 61: 721–759

Theißen G, Saedler H (1995) MADS-box genes in plant ontogeny and phylogeny: Haeckel's "biogenetic law" revisited. Curr Opin Genet Dev 5: 628–639

Thiney Y (1972) Morphologie et cytochimie ultrastructurale de l' oscule d' *Hippospongia communis* LMK et de sa régénération. Thèse, Univ Claude Bernard, pp 1–63

Tsai J, Douglas MG (1996) A conserved HPD sequence of the J-domain is necessary for YDJ1 stimulation of hsp70 ATPase activity at a site distinct from substrate binding. J Biol Chem 272: 9347–9354

Valentine JM (1994) Late precambrian vilaterians: grades and clades. Proc Natl Acad Sci USA 91: 6751–6757

Valentine JW, Collins AG, Meyer CP (1993) Morphological complexity increase in metazoans. Palaeobiology 20: 131–142

Valentine JW, Erwin DH, Jablonski D (1996) Developmental evolution of metazoan body plans: the fossil evidence. Dev Biol 173: 373–381

Vogel S (1977) Current-induced flow through living sponges in nature. Proc Natl Acad Sci USA 74: 2069–2071

Wagner-Hülsmann C, Bachinski N, Diehl-Seifert B, Blumbach B, Steffen R, Pancer Z, Müller WEG (1996) A galectin links the aggregation factor to cells in the sponge [*Geodia cydonium*] system. Glycobiology 6: 785–793

Wall D, Zylicz M, Georgopoulos C (1995) The conserved G/F motif of the DnaJ chaperone is necessary for the activation of the substrate binding properties of the DnaK chaperone. J Biol Chem 270: 2139–2144

Walter M (1996) Old fossils could be fractal frauds. Nature 383: 385–386

Walter L, Rauh F, Günther E (1994) Comparative analysis of the three major histocompatibility complex-linked Hsp70 genes of the rat. Immunogenetics 40: 325–330

Weinbaum G, Burger MM (1973) A two-component system for surface guided reassociation of animal cells. Nature 244: 510–512

Wijngaard PL, Metzelaar MJ, MacHugh ND, Morrison WI, Clevers HC (1992) Molecular characterization of the WC1 antigen expressed specifically on bovine CD4-CD8-gamma delta T lymphocytes. J Immunol 149: 3273–3277

Williams AF, Barclay AN (1988) The immunoglobulin superfamily - domains for cell surface recognition. Annu Rev Immunol 6: 381–405

Willmer P (1994) Invertebrate relationships. Cambridge University Press, Cambridge

Wistow G (1993) Lens crystallins: gene recruitment and evolutionary dynamism. Trends Biochem Sci 18: 301–306

Wistow GJ, Piatigorsky J (1988) Lens crystallins: the evolution and expression of proteins for a highly specialized tissue. Annu Rev Biochem 57: 479–504

Woese CR (1987) Bacterial evolution. Microbiol Rev 51: 221–271

Woese CR, Achenbach L, Rouviere P, Mandelco L (1991) Archaeal phylogeny: reexamination of the phylogenetic position of *Archaeoglobus fulgidus* in light of certain composition-induced artifacts. Syst Appl Microbiol 14: 364–371

Yarfitz S, Hurley JB (1994) Transduction mechanisms of vertebrate and invertebrate photoreceptors. J Biol Chem 269: 14329–14332

Yu Q, Reichert M, Brousseau T, Cleute Y, Burny A, Kettmann R (1990) Sequence of bovine CD5. Nucl Acids Res 18: 5296

Zipfel PF, Skerka C (1994) Complement factor H and related proteins: an expanding family of complement-regulatory proteins? Immunol Today 15: 121–126

Homeobox Genes in the Freshwater Sponge *Ephydatia fluviatilis*

Makiko Seimiya[1,2], Morihiro Naito[1], Yoko Watanabe[2], and Yoshikazu Kurosawa[1]

1 Introduction

Homeobox genes, which contain a conserved 180 base pair (bp) sequence known as a homeobox, were first identified in segmented metazoans (McGinnis et al. 1984a; Scott and Weiner 1984). Subsequently, many homeobox genes have been identified in nonsegmented metazoans, such as nematodes, Plathyhelminthes and Cnidaria (Bürglin 1994). Sponges are the most primitive extant metazoans and the simplest living multicellular animals. Each consists of several types of cells, but none has any distinctive organs, a specialized digestive apparatus, or an excretory system. Sponges also lack a nervous system. Since sponges seem to have a body plan that is different from that of other metazoans, they were originally considered to be a phylum that evolved independently of the major pathway of evolution in the animal kingdom (Hyman 1951). In other words, sponges and other animals were believed to have derived paraphyletically from different protists. However, this concept has recently been challenged since many lines of evidence, including sequences of 18 S rRNA, support the hypothesis that sponges belong to a phylum that has evolved in the same way as other metazoan phyla (Field et al. 1988; Lake 1990; Christen et al. 1991; Wainright et al. 1993), and, moreover, that metazoans, including sponges, form a monophyletic taxon.

The homeodomains encoded by homeoboxes can bind to DNA in a sequence-specific manner (Hayashi and Scott 1990; Pabo and Sauer 1992), and they can be found in large numbers of transcription factors from various organisms (Bürglin 1994). Homeobox genes have been isolated not only from animals but also from plants and fungi (Bürglin 1994). Thus, homeodomains represent one of the major types of structural motif that can bind to DNA.

From the similarities among their sequences, homeobox genes have been divided into many classes (Scott et al. 1989; Dessain and McGinnis 1993; Kappen et al. 1993; Bürglin 1994). Within each class of homeodomain, the

[1] Institute for Comprehensive Medical Science, Fujita Health University, Toyoake, Aichi 470-11, Japan
[2] Department of Biology, Faculty of Science, Ochanomizu University, 2-1-1 Otsuka, Tokyo 112, Japan

Progress in Molecular and Subcellular Biology, Vol. 19
W.E.G. Müller (Ed.)

homeodomains from *Drosophila* and mammals are more than 80% identical in terms of amino acid residues, and it seems likely, therefore, that representatives of each class of homeodomains became established before the divergence of arthropods and vertebrates in evolution (Akam 1989; Kenyon and Wang 1991; McGinnis and Krumlauf 1992). The homologous genes of *Drosophila* and mammals are not only structurally similar to one another, but they are also functionally interchangeable in many cases (McGinnis et al. 1990; Malicki et al. 1992; Bachiller et al. 1994; Quiring et al. 1994; Frasch et al. 1995).

In this chapter, we summarize the available evidence that the homeobox genes identified in freshwater sponges (Seimiya et al. 1994, 1997) have the same characteristics as members of the different classes of homeobox genes that have been identified in higher organisms. We also discuss general principles that appear to be applicable to the observed conservation and diversification of homeobox genes during animal evolution.

2 Classification of Homeobox Genes

2.1 Characteristics of Homeodomains

When we compare the amino acid sequences of entire proteins encoded by functionally identical homeobox genes between phylogenically distant animals, such as mammals and *Drosophila*, it is apparent that the homologous regions are restricted to the homeodomains, with the exception of some other functional motifs (Bürglin 1994). Thus, classification of homeobox genes can be based primarily on the amino acid sequences of the respective homeodomains (Scott et al. 1989; Dessain and McGinnis 1993; Kappen et al. 1993; Bürglin 1994).

The three-dimensional structure of homeodomains and that of complexes formed between a homeodomain and its target DNA have been determined by

Fig. 1A–C. A summary of the three-dimensional structures of homeodomains, as analyzed by X-ray crystallography and NMR spectroscopy. Seven kinds of homeodomains were analyzed. The results for the ftz homeodomain (Qian et al. 1994) are not included. **A** Three α-helical portions are *boxed*. *Open circles* indicate the residues that are involved in formation of a hydrophobic core. *Closed triangles* indicate the residues whose side chains make direct contact with the target DNA. **B** The overall structure of the complex of a homeodomain with DNA. *1*, *2*, and *3* indicate the α-helices. **C** The location of each amino acid residue is indicated, with emphasis on the distribution of residues in the α-helical regions. The *white numbers on a black background* indicate the residues that are involved in formation of a hydrophobic core. The *shadowed numbers* indicate the residues involved in direct contacts with DNA. Strongly conserved residues are *circled with a bold line*. The data were taken from the following sources: Antp (Qian et al. 1989; Otting et al. 1990; Billeter et al. 1993); En (Kissinger et al. 1990); MATα2 (Wolberger et al. 1991); LF-B1 (Cesk et al. 1993; Leiting et al. 1993); Oct-1 (Klemm et al. 1994); and Eve (Hirsch and Aggarwel 1995)

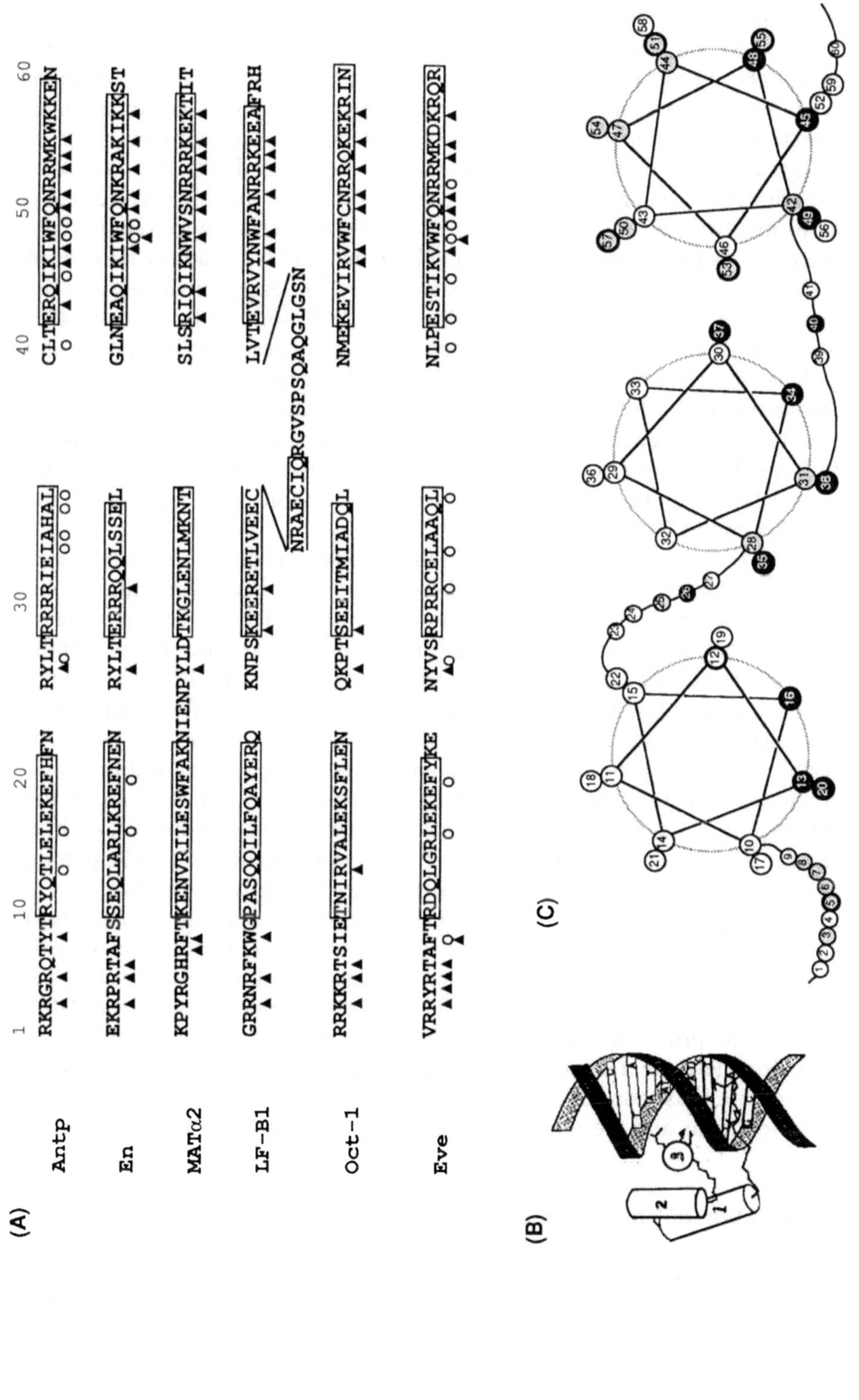
(A)
1 10 20 30 40 50 60
Antp RKRGRQTYTRYQTLELEKEFHFN RYLTRRRRIEIAHAL CLTERQIKIWFQNRRMKWKKEN
En EKRPRTAFSSEQLARLKREFNEN RYLTERRRQQLSSEL GLNEAQIKIWFQNKRAKIKKST
MATα2 KPYRGHRFTKENVRILESWFAKNIENPYLDTKGLENLMKNT SLSRIQIKNWVSNRRRKEKTIT
LF-B1 GRRNRFKWGPASQQILFQAYERQ KNPSKEERETLVEEC NRAECIQRGVSPSQAQGLGSN LVTEVRVYNWFANRRKEEAFRH
Oct-1 RRKKRTSIETNIRVALEKSFLEN QKPTSEEITMIADQL NMEKEVIRVWFCNRRQKEKRIN
Eve VRRYRTAFTRDQLGRLEKEFYKE NYVSRPRRCELAAQL NLPESTIKVWFQNRRMKDKRQR
(B)
(C)

Fig. 1A–C.

X-ray analysis of crystals and by NMR analysis of complexes in solution (Kissinger et al. 1990; Otting et al. 1990). Figure 1 shows a summary of the results. Seven different kinds of homeodomains have been analyzed (Kissinger et al. 1990; Otting et al. 1990; Wolberger et al. 1991; Cestka et al. 1993; Klemm et al. 1994; Qian et al. 1994; Hirsch and Aggarwal 1995). The fundamental structures of all homeodomains appear to be similar. Each contains three α-helices and an extended amino-terminal arm. Helix 1 and helix 2 are apposed to each other in an antiparallel arrangement. Helix 3 is roughly perpendicular to the first two helices (Fig. 1B). Gehring's and Wuthrich's groups reported the presence of a kink at the point between the 52nd and 53rd residues in the third helix of the Antp homeodomain (Qian et al. 1989). At least 11 residues, namely, those at the 13th, 16th, 26th, 34th, 35th, 37th, 38th, 40th, 45th, 48th and 49th positions, appear to form a hydrophobic core, with resultant formation of a globular structure (Qian et al. 1989). In general, the hydrophobic amino acids that are involved in the formation of the core structure of a protein tend to be strongly conserved, and such is the case for the homeodomains. Leucine at the 16th position, leucine at the 40th position, tryptophan at the 48th position, and phenylalanine at the 49th position are strongly conserved among all the homeodomains (Bürglin 1994).

There seems to be general agreement that helix 3 fits directly into the major groove of the target DNA and that the side chains of the hydrophilic amino acids, located on the surface of the homeodomain, make extensive contacts with the DNA. Moreover, the extended amino-terminal arm fits into the minor groove of the DNA, where it makes several contacts with the DNA. Among the amino acid residues that make contact with the DNA, some make contacts with the sugar-phosphate backbone, for example, tyrosine at the 25th position, arginine at the 31st position, tryptophan at the 48th position, and the basic amino acids at the 53rd, 55th, and 57th positions. Several residues in the amino-terminal arm and in helix 3 make sequence-specific contacts, such as van der Waals' contacts and hydrogen bonds, with specific bases. The precise positions of the amino acid residues that make contacts with the DNA appear to differ slightly among homeodomains, as summarized in Fig. 1. The reason for the high degree of conservation of the amino acid sequence of each homeodomain for which homologues have been found in various organisms could be that it is critical for amino acid residues that make contact with DNA to be in the proper specific configuration. Therefore, to classify homeobox genes from lower metazoans into the classes that have been identified in higher animals, we must consider both the overall similarities among sequences of homeodomains and the specific residues that make contacts with the DNA.

2.2 Classification of Homeobox Genes in *Drosophila*

More than 70 homeobox genes have been isolated from *Drosophila* to date (Dessain and McGinnis 1993; Bürglin 1994). Although two or three times this

Table 1. Comparison of the amino acid sequences of homeodomains encoded by homeobox genes of Drosophila and sponge. Fifty-eight genes from Drosophila and five genes from sponge are compared. The numerical values indicate the percentage of homology for the 60-amino-acid sequences of the homeodomains. To emphasize the different degrees of homology, values are shown in boxes of different colors (from red via orange, yellow, and green to blue at intervals of 10%). The references for the amino acid sequences encoded by all the homeobox genes can be found in reports by Dessain and McGinnis (1993) and Bürglin (1994), with the exception of the following: lbe (Jagla et al. 1994), Otp (Simeone et al. 1994), and al (Schneitz et al. 1993). The Ey gene, the Pax-6 homologue in Drosophila (Quiring et al. 1994), is not included

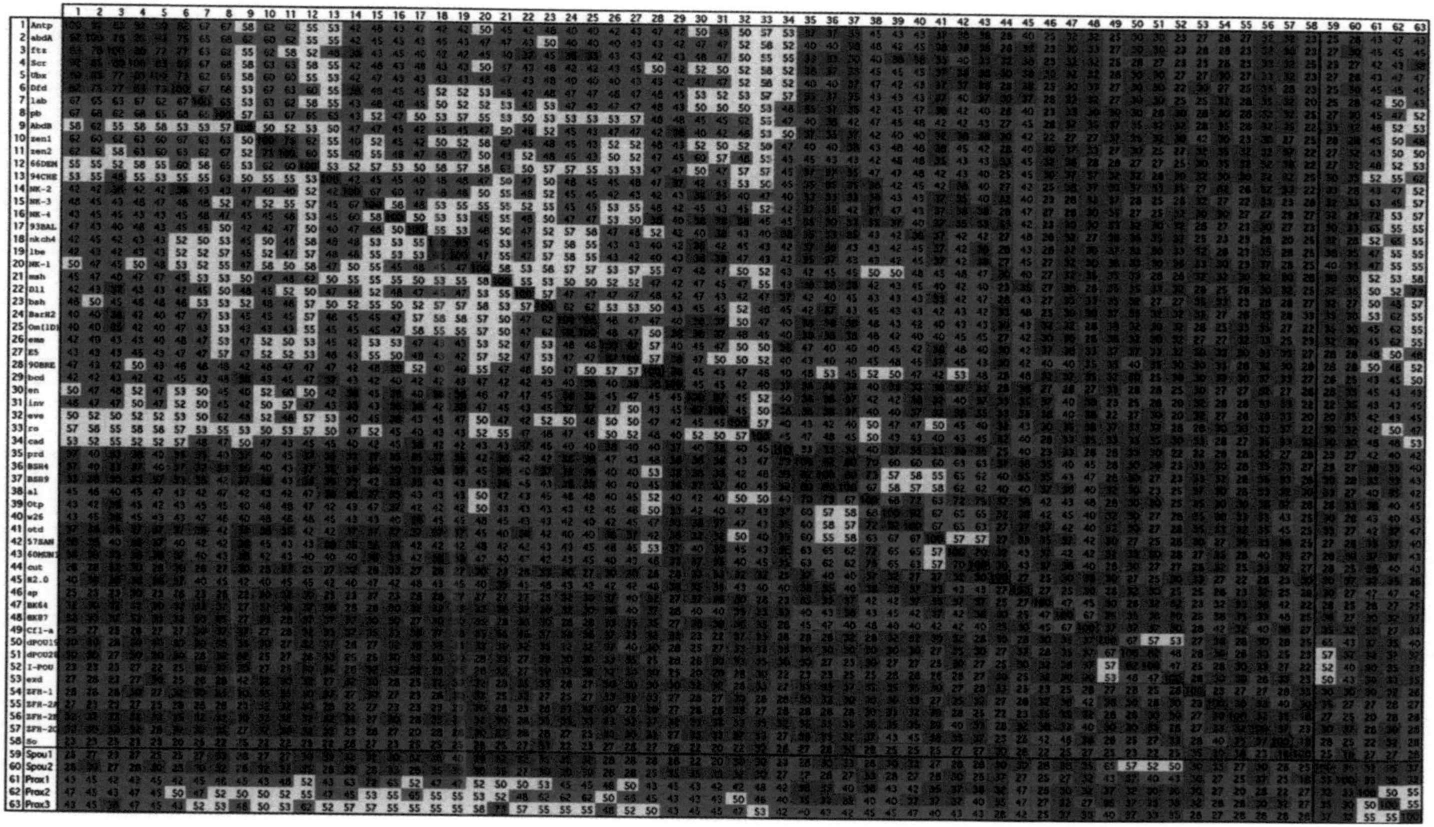

	1	2	3	4	5	6	7	8	9	10	11	12	13	14	15	16	17	18	19	20	21	22	23	24	25	26	27	28	29	30	31	32	33	34	35	36	37	38	39	40	41	42	43	44	45	46	47	48	49	50	51	52	53	54	55	56	57	58	59	60	61	62	63
1 Antp	[illegible]	[illegible]	[illegible]	[illegible]	[illegible]	[illegible]	[illegible]	[illegible]	[illegible]	[illegible]	[illegible]	[illegible]	[illegible]	[illegible]	[illegible]	[illegible]	[illegible]	[illegible]	[illegible]	[illegible]	[illegible]	[illegible]	[illegible]	[illegible]	[illegible]	[illegible]	[illegible]	[illegible]	[illegible]	[illegible]	[illegible]	[illegible]	[illegible]	[illegible]	[illegible]	[illegible]	[illegible]	[illegible]	[illegible]	[illegible]	[illegible]	[illegible]	[illegible]	[illegible]	[illegible]	[illegible]	[illegible]	[illegible]	[illegible]	[illegible]	[illegible]	[illegible]	[illegible]	[illegible]	[illegible]	[illegible]	[illegible]	[illegible]	[illegible]	[illegible]	[illegible]	[illegible]	[illegible]
2 abdA	[illegible]	[illegible]	[illegible]	[illegible]	[illegible]	[illegible]	[illegible]	[illegible]	[illegible]	[illegible]	[illegible]	[illegible]	[illegible]	[illegible]	[illegible]	[illegible]	[illegible]	[illegible]	[illegible]	[illegible]	[illegible]	[illegible]	[illegible]	[illegible]	[illegible]	[illegible]	[illegible]	[illegible]	[illegible]	[illegible]	[illegible]	[illegible]	[illegible]	[illegible]	[illegible]	[illegible]	[illegible]	[illegible]	[illegible]	[illegible]	[illegible]	[illegible]	[illegible]	[illegible]	[illegible]	[illegible]	[illegible]	[illegible]	[illegible]	[illegible]	[illegible]	[illegible]	[illegible]	[illegible]	[illegible]	[illegible]	[illegible]	[illegible]	[illegible]	[illegible]	[illegible]	[illegible]	[illegible]
3 ftz	[illegible]	[illegible]	[illegible]	[illegible]	[illegible]	[illegible]	[illegible]	[illegible]	[illegible]	[illegible]	[illegible]	[illegible]	[illegible]	[illegible]	[illegible]	[illegible]	[illegible]	[illegible]	[illegible]	[illegible]	[illegible]	[illegible]	[illegible]	[illegible]	[illegible]	[illegible]	[illegible]	[illegible]	[illegible]	[illegible]	[illegible]	[illegible]	[illegible]	[illegible]	[illegible]	[illegible]	[illegible]	[illegible]	[illegible]	[illegible]	[illegible]	[illegible]	[illegible]	[illegible]	[illegible]	[illegible]	[illegible]	[illegible]	[illegible]	[illegible]	[illegible]	[illegible]	[illegible]	[illegible]	[illegible]	[illegible]	[illegible]	[illegible]	[illegible]	[illegible]	[illegible]	[illegible]	[illegible]
4 Scr	[illegible]	[illegible]	[illegible]	[illegible]	[illegible]	[illegible]	[illegible]	[illegible]	[illegible]	[illegible]	[illegible]	[illegible]	[illegible]	[illegible]	[illegible]	[illegible]	[illegible]	[illegible]	[illegible]	[illegible]	[illegible]	[illegible]	[illegible]	[illegible]	[illegible]	[illegible]	[illegible]	[illegible]	[illegible]	[illegible]	[illegible]	[illegible]	[illegible]	[illegible]	[illegible]	[illegible]	[illegible]	[illegible]	[illegible]	[illegible]	[illegible]	[illegible]	[illegible]	[illegible]	[illegible]	[illegible]	[illegible]	[illegible]	[illegible]	[illegible]	[illegible]	[illegible]	[illegible]	[illegible]	[illegible]	[illegible]	[illegible]	[illegible]	[illegible]	[illegible]	[illegible]	[illegible]	[illegible]
5 Ubx	[illegible]	[illegible]	[illegible]	[illegible]	[illegible]	[illegible]	[illegible]	[illegible]	[illegible]	[illegible]	[illegible]	[illegible]	[illegible]	[illegible]	[illegible]	[illegible]	[illegible]	[illegible]	[illegible]	[illegible]	[illegible]	[illegible]	[illegible]	[illegible]	[illegible]	[illegible]	[illegible]	[illegible]	[illegible]	[illegible]	[illegible]	[illegible]	[illegible]	[illegible]	[illegible]	[illegible]	[illegible]	[illegible]	[illegible]	[illegible]	[illegible]	[illegible]	[illegible]	[illegible]	[illegible]	[illegible]	[illegible]	[illegible]	[illegible]	[illegible]	[illegible]	[illegible]	[illegible]	[illegible]	[illegible]	[illegible]	[illegible]	[illegible]	[illegible]	[illegible]	[illegible]	[illegible]	[illegible]
6 Dfd	[illegible]	[illegible]	[illegible]	[illegible]	[illegible]	[illegible]	[illegible]	[illegible]	[illegible]	[illegible]	[illegible]	[illegible]	[illegible]	[illegible]	[illegible]	[illegible]	[illegible]	[illegible]	[illegible]	[illegible]	[illegible]	[illegible]	[illegible]	[illegible]	[illegible]	[illegible]	[illegible]	[illegible]	[illegible]	[illegible]	[illegible]	[illegible]	[illegible]	[illegible]	[illegible]	[illegible]	[illegible]	[illegible]	[illegible]	[illegible]	[illegible]	[illegible]	[illegible]	[illegible]	[illegible]	[illegible]	[illegible]	[illegible]	[illegible]	[illegible]	[illegible]	[illegible]	[illegible]	[illegible]	[illegible]	[illegible]	[illegible]	[illegible]	[illegible]	[illegible]	[illegible]	[illegible]	[illegible]
7 lab	[illegible]	[illegible]	[illegible]	[illegible]	[illegible]	[illegible]	[illegible]	[illegible]	[illegible]	[illegible]	[illegible]	[illegible]	[illegible]	[illegible]	[illegible]	[illegible]	[illegible]	[illegible]	[illegible]	[illegible]	[illegible]	[illegible]	[illegible]	[illegible]	[illegible]	[illegible]	[illegible]	[illegible]	[illegible]	[illegible]	[illegible]	[illegible]	[illegible]	[illegible]	[illegible]	[illegible]	[illegible]	[illegible]	[illegible]	[illegible]	[illegible]	[illegible]	[illegible]	[illegible]	[illegible]	[illegible]	[illegible]	[illegible]	[illegible]	[illegible]	[illegible]	[illegible]	[illegible]	[illegible]	[illegible]	[illegible]	[illegible]	[illegible]	[illegible]	[illegible]	[illegible]	[illegible]	[illegible]
8 pb	[illegible]	[illegible]	[illegible]	[illegible]	[illegible]	[illegible]	[illegible]	[illegible]	[illegible]	[illegible]	[illegible]	[illegible]	[illegible]	[illegible]	[illegible]	[illegible]	[illegible]	[illegible]	[illegible]	[illegible]	[illegible]	[illegible]	[illegible]	[illegible]	[illegible]	[illegible]	[illegible]	[illegible]	[illegible]	[illegible]	[illegible]	[illegible]	[illegible]	[illegible]	[illegible]	[illegible]	[illegible]	[illegible]	[illegible]	[illegible]	[illegible]	[illegible]	[illegible]	[illegible]	[illegible]	[illegible]	[illegible]	[illegible]	[illegible]	[illegible]	[illegible]	[illegible]	[illegible]	[illegible]	[illegible]	[illegible]	[illegible]	[illegible]	[illegible]	[illegible]	[illegible]	[illegible]	[illegible]
9 AbdB	[illegible]	[illegible]	[illegible]	[illegible]	[illegible]	[illegible]	[illegible]	[illegible]	[illegible]	[illegible]	[illegible]	[illegible]	[illegible]	[illegible]	[illegible]	[illegible]	[illegible]	[illegible]	[illegible]	[illegible]	[illegible]	[illegible]	[illegible]	[illegible]	[illegible]	[illegible]	[illegible]	[illegible]	[illegible]	[illegible]	[illegible]	[illegible]	[illegible]	[illegible]	[illegible]	[illegible]	[illegible]	[illegible]	[illegible]	[illegible]	[illegible]	[illegible]	[illegible]	[illegible]	[illegible]	[illegible]	[illegible]	[illegible]	[illegible]	[illegible]	[illegible]	[illegible]	[illegible]	[illegible]	[illegible]	[illegible]	[illegible]	[illegible]	[illegible]	[illegible]	[illegible]	[illegible]	[illegible]
10 zen1	[illegible]	[illegible]	[illegible]	[illegible]	[illegible]	[illegible]	[illegible]	[illegible]	[illegible]	[illegible]	[illegible]	[illegible]	[illegible]	[illegible]	[illegible]	[illegible]	[illegible]	[illegible]	[illegible]	[illegible]	[illegible]	[illegible]	[illegible]	[illegible]	[illegible]	[illegible]	[illegible]	[illegible]	[illegible]	[illegible]	[illegible]	[illegible]	[illegible]	[illegible]	[illegible]	[illegible]	[illegible]	[illegible]	[illegible]	[illegible]	[illegible]	[illegible]	[illegible]	[illegible]	[illegible]	[illegible]	[illegible]	[illegible]	[illegible]	[illegible]	[illegible]	[illegible]	[illegible]	[illegible]	[illegible]	[illegible]	[illegible]	[illegible]	[illegible]	[illegible]	[illegible]	[illegible]	[illegible]
11 zen2	[illegible]	[illegible]	[illegible]	[illegible]	[illegible]	[illegible]	[illegible]	[illegible]	[illegible]	[illegible]	[illegible]	[illegible]	[illegible]	[illegible]	[illegible]	[illegible]	[illegible]	[illegible]	[illegible]	[illegible]	[illegible]	[illegible]	[illegible]	[illegible]	[illegible]	[illegible]	[illegible]	[illegible]	[illegible]	[illegible]	[illegible]	[illegible]	[illegible]	[illegible]	[illegible]	[illegible]	[illegible]	[illegible]	[illegible]	[illegible]	[illegible]	[illegible]	[illegible]	[illegible]	[illegible]	[illegible]	[illegible]	[illegible]	[illegible]	[illegible]	[illegible]	[illegible]	[illegible]	[illegible]	[illegible]	[illegible]	[illegible]	[illegible]	[illegible]	[illegible]	[illegible]	[illegible]	[illegible]
12 66DEN	[illegible]	[illegible]	[illegible]	[illegible]	[illegible]	[illegible]	[illegible]	[illegible]	[illegible]	[illegible]	[illegible]	[illegible]	[illegible]	[illegible]	[illegible]	[illegible]	[illegible]	[illegible]	[illegible]	[illegible]	[illegible]	[illegible]	[illegible]	[illegible]	[illegible]	[illegible]	[illegible]	[illegible]	[illegible]	[illegible]	[illegible]	[illegible]	[illegible]	[illegible]	[illegible]	[illegible]	[illegible]	[illegible]	[illegible]	[illegible]	[illegible]	[illegible]	[illegible]	[illegible]	[illegible]	[illegible]	[illegible]	[illegible]	[illegible]	[illegible]	[illegible]	[illegible]	[illegible]	[illegible]	[illegible]	[illegible]	[illegible]	[illegible]	[illegible]	[illegible]	[illegible]	[illegible]	[illegible]
13 94CHE	[illegible]	[illegible]	[illegible]	[illegible]	[illegible]	[illegible]	[illegible]	[illegible]	[illegible]	[illegible]	[illegible]	[illegible]	[illegible]	[illegible]	[illegible]	[illegible]	[illegible]	[illegible]	[illegible]	[illegible]	[illegible]	[illegible]	[illegible]	[illegible]	[illegible]	[illegible]	[illegible]	[illegible]	[illegible]	[illegible]	[illegible]	[illegible]	[illegible]	[illegible]	[illegible]	[illegible]	[illegible]	[illegible]	[illegible]	[illegible]	[illegible]	[illegible]	[illegible]	[illegible]	[illegible]	[illegible]	[illegible]	[illegible]	[illegible]	[illegible]	[illegible]	[illegible]	[illegible]	[illegible]	[illegible]	[illegible]	[illegible]	[illegible]	[illegible]	[illegible]	[illegible]	[illegible]	[illegible]
14 NK-2	[illegible]	[illegible]	[illegible]	[illegible]	[illegible]	[illegible]	[illegible]	[illegible]	[illegible]	[illegible]	[illegible]	[illegible]	[illegible]	[illegible]	[illegible]	[illegible]	[illegible]	[illegible]	[illegible]	[illegible]	[illegible]	[illegible]	[illegible]	[illegible]	[illegible]	[illegible]	[illegible]	[illegible]	[illegible]	[illegible]	[illegible]	[illegible]	[illegible]	[illegible]	[illegible]	[illegible]	[illegible]	[illegible]	[illegible]	[illegible]	[illegible]	[illegible]	[illegible]	[illegible]	[illegible]	[illegible]	[illegible]	[illegible]	[illegible]	[illegible]	[illegible]	[illegible]	[illegible]	[illegible]	[illegible]	[illegible]	[illegible]	[illegible]	[illegible]	[illegible]	[illegible]	[illegible]	[illegible]
15 NK-3	[illegible]	[illegible]	[illegible]	[illegible]	[illegible]	[illegible]	[illegible]	[illegible]	[illegible]	[illegible]	[illegible]	[illegible]	[illegible]	[illegible]	[illegible]	[illegible]	[illegible]	[illegible]	[illegible]	[illegible]	[illegible]	[illegible]	[illegible]	[illegible]	[illegible]	[illegible]	[illegible]	[illegible]	[illegible]	[illegible]	[illegible]	[illegible]	[illegible]	[illegible]	[illegible]	[illegible]	[illegible]	[illegible]	[illegible]	[illegible]	[illegible]	[illegible]	[illegible]	[illegible]	[illegible]	[illegible]	[illegible]	[illegible]	[illegible]	[illegible]	[illegible]	[illegible]	[illegible]	[illegible]	[illegible]	[illegible]	[illegible]	[illegible]	[illegible]	[illegible]	[illegible]	[illegible]	[illegible]
16 NK-4	[illegible]	[illegible]	[illegible]	[illegible]	[illegible]	[illegible]	[illegible]	[illegible]	[illegible]	[illegible]	[illegible]	[illegible]	[illegible]	[illegible]	[illegible]	[illegible]	[illegible]	[illegible]	[illegible]	[illegible]	[illegible]	[illegible]	[illegible]	[illegible]	[illegible]	[illegible]	[illegible]	[illegible]	[illegible]	[illegible]	[illegible]	[illegible]	[illegible]	[illegible]	[illegible]	[illegible]	[illegible]	[illegible]	[illegible]	[illegible]	[illegible]	[illegible]	[illegible]	[illegible]	[illegible]	[illegible]	[illegible]	[illegible]	[illegible]	[illegible]	[illegible]	[illegible]	[illegible]	[illegible]	[illegible]	[illegible]	[illegible]	[illegible]	[illegible]	[illegible]	[illegible]	[illegible]	[illegible]
17 93BAL	[illegible]	[illegible]	[illegible]	[illegible]	[illegible]	[illegible]	[illegible]	[illegible]	[illegible]	[illegible]	[illegible]	[illegible]	[illegible]	[illegible]	[illegible]	[illegible]	[illegible]	[illegible]	[illegible]	[illegible]	[illegible]	[illegible]	[illegible]	[illegible]	[illegible]	[illegible]	[illegible]	[illegible]	[illegible]	[illegible]	[illegible]	[illegible]	[illegible]	[illegible]	[illegible]	[illegible]	[illegible]	[illegible]	[illegible]	[illegible]	[illegible]	[illegible]	[illegible]	[illegible]	[illegible]	[illegible]	[illegible]	[illegible]	[illegible]	[illegible]	[illegible]	[illegible]	[illegible]	[illegible]	[illegible]	[illegible]	[illegible]	[illegible]	[illegible]	[illegible]	[illegible]	[illegible]	[illegible]
18 nkch4	[illegible]	[illegible]	[illegible]	[illegible]	[illegible]	[illegible]	[illegible]	[illegible]	[illegible]	[illegible]	[illegible]	[illegible]	[illegible]	[illegible]	[illegible]	[illegible]	[illegible]	[illegible]	[illegible]	[illegible]	[illegible]	[illegible]	[illegible]	[illegible]	[illegible]	[illegible]	[illegible]	[illegible]	[illegible]	[illegible]	[illegible]	[illegible]	[illegible]	[illegible]	[illegible]	[illegible]	[illegible]	[illegible]	[illegible]	[illegible]	[illegible]	[illegible]	[illegible]	[illegible]	[illegible]	[illegible]	[illegible]	[illegible]	[illegible]	[illegible]	[illegible]	[illegible]	[illegible]	[illegible]	[illegible]	[illegible]	[illegible]	[illegible]	[illegible]	[illegible]	[illegible]	[illegible]	[illegible]
19 lbe	[illegible]	[illegible]	[illegible]	[illegible]	[illegible]	[illegible]	[illegible]	[illegible]	[illegible]	[illegible]	[illegible]	[illegible]	[illegible]	[illegible]	[illegible]	[illegible]	[illegible]	[illegible]	[illegible]	[illegible]	[illegible]	[illegible]	[illegible]	[illegible]	[illegible]	[illegible]	[illegible]	[illegible]	[illegible]	[illegible]	[illegible]	[illegible]	[illegible]	[illegible]	[illegible]	[illegible]	[illegible]	[illegible]	[illegible]	[illegible]	[illegible]	[illegible]	[illegible]	[illegible]	[illegible]	[illegible]	[illegible]	[illegible]	[illegible]	[illegible]	[illegible]	[illegible]	[illegible]	[illegible]	[illegible]	[illegible]	[illegible]	[illegible]	[illegible]	[illegible]	[illegible]	[illegible]	[illegible]
20 NK-1	[illegible]	[illegible]	[illegible]	[illegible]	[illegible]	[illegible]	[illegible]	[illegible]	[illegible]	[illegible]	[illegible]	[illegible]	[illegible]	[illegible]	[illegible]	[illegible]	[illegible]	[illegible]	[illegible]	[illegible]	[illegible]	[illegible]	[illegible]	[illegible]	[illegible]	[illegible]	[illegible]	[illegible]	[illegible]	[illegible]	[illegible]	[illegible]	[illegible]	[illegible]	[illegible]	[illegible]	[illegible]	[illegible]	[illegible]	[illegible]	[illegible]	[illegible]	[illegible]	[illegible]	[illegible]	[illegible]	[illegible]	[illegible]	[illegible]	[illegible]	[illegible]	[illegible]	[illegible]	[illegible]	[illegible]	[illegible]	[illegible]	[illegible]	[illegible]	[illegible]	[illegible]	[illegible]	[illegible]
21 msh	[illegible]	[illegible]	[illegible]	[illegible]	[illegible]	[illegible]	[illegible]	[illegible]	[illegible]	[illegible]	[illegible]	[illegible]	[illegible]	[illegible]	[illegible]	[illegible]	[illegible]	[illegible]	[illegible]	[illegible]	[illegible]	[illegible]	[illegible]	[illegible]	[illegible]	[illegible]	[illegible]	[illegible]	[illegible]	[illegible]	[illegible]	[illegible]	[illegible]	[illegible]	[illegible]	[illegible]	[illegible]	[illegible]	[illegible]	[illegible]	[illegible]	[illegible]	[illegible]	[illegible]	[illegible]	[illegible]	[illegible]	[illegible]	[illegible]	[illegible]	[illegible]	[illegible]	[illegible]	[illegible]	[illegible]	[illegible]	[illegible]	[illegible]	[illegible]	[illegible]	[illegible]	[illegible]	[illegible]
22 Dll	[illegible]	[illegible]	[illegible]	[illegible]	[illegible]	[illegible]	[illegible]	[illegible]	[illegible]	[illegible]	[illegible]	[illegible]	[illegible]	[illegible]	[illegible]	[illegible]	[illegible]	[illegible]	[illegible]	[illegible]	[illegible]	[illegible]	[illegible]	[illegible]	[illegible]	[illegible]	[illegible]	[illegible]	[illegible]	[illegible]	[illegible]	[illegible]	[illegible]	[illegible]	[illegible]	[illegible]	[illegible]	[illegible]	[illegible]	[illegible]	[illegible]	[illegible]	[illegible]	[illegible]	[illegible]	[illegible]	[illegible]	[illegible]	[illegible]	[illegible]	[illegible]	[illegible]	[illegible]	[illegible]	[illegible]	[illegible]	[illegible]	[illegible]	[illegible]	[illegible]	[illegible]	[illegible]	[illegible]
23 bsh	[illegible]	[illegible]	[illegible]	[illegible]	[illegible]	[illegible]	[illegible]	[illegible]	[illegible]	[illegible]	[illegible]	[illegible]	[illegible]	[illegible]	[illegible]	[illegible]	[illegible]	[illegible]	[illegible]	[illegible]	[illegible]	[illegible]	[illegible]	[illegible]	[illegible]	[illegible]	[illegible]	[illegible]	[illegible]	[illegible]	[illegible]	[illegible]	[illegible]	[illegible]	[illegible]	[illegible]	[illegible]	[illegible]	[illegible]	[illegible]	[illegible]	[illegible]	[illegible]	[illegible]	[illegible]	[illegible]	[illegible]	[illegible]	[illegible]	[illegible]	[illegible]	[illegible]	[illegible]	[illegible]	[illegible]	[illegible]	[illegible]	[illegible]	[illegible]	[illegible]	[illegible]	[illegible]	[illegible]
24 BarH2	[illegible]	[illegible]	[illegible]	[illegible]	[illegible]	[illegible]	[illegible]	[illegible]	[illegible]	[illegible]	[illegible]	[illegible]	[illegible]	[illegible]	[illegible]	[illegible]	[illegible]	[illegible]	[illegible]	[illegible]	[illegible]	[illegible]	[illegible]	[illegible]	[illegible]	[illegible]	[illegible]	[illegible]	[illegible]	[illegible]	[illegible]	[illegible]	[illegible]	[illegible]	[illegible]	[illegible]	[illegible]	[illegible]	[illegible]	[illegible]	[illegible]	[illegible]	[illegible]	[illegible]	[illegible]	[illegible]	[illegible]	[illegible]	[illegible]	[illegible]	[illegible]	[illegible]	[illegible]	[illegible]	[illegible]	[illegible]	[illegible]	[illegible]	[illegible]	[illegible]	[illegible]	[illegible]	[illegible]
25 Om(1D)	[illegible]	[illegible]	[illegible]	[illegible]	[illegible]	[illegible]	[illegible]	[illegible]	[illegible]	[illegible]	[illegible]	[illegible]	[illegible]	[illegible]	[illegible]	[illegible]	[illegible]	[illegible]	[illegible]	[illegible]	[illegible]	[illegible]	[illegible]	[illegible]	[illegible]	[illegible]	[illegible]	[illegible]	[illegible]	[illegible]	[illegible]	[illegible]	[illegible]	[illegible]	[illegible]	[illegible]	[illegible]	[illegible]	[illegible]	[illegible]	[illegible]	[illegible]	[illegible]	[illegible]	[illegible]	[illegible]	[illegible]	[illegible]	[illegible]	[illegible]	[illegible]	[illegible]	[illegible]	[illegible]	[illegible]	[illegible]	[illegible]	[illegible]	[illegible]	[illegible]	[illegible]	[illegible]	[illegible]
26 ems	[illegible]	[illegible]	[illegible]	[illegible]	[illegible]	[illegible]	[illegible]	[illegible]	[illegible]	[illegible]	[illegible]	[illegible]	[illegible]	[illegible]	[illegible]	[illegible]	[illegible]	[illegible]	[illegible]	[illegible]	[illegible]	[illegible]	[illegible]	[illegible]	[illegible]	[illegible]	[illegible]	[illegible]	[illegible]	[illegible]	[illegible]	[illegible]	[illegible]	[illegible]	[illegible]	[illegible]	[illegible]	[illegible]	[illegible]	[illegible]	[illegible]	[illegible]	[illegible]	[illegible]	[illegible]	[illegible]	[illegible]	[illegible]	[illegible]	[illegible]	[illegible]	[illegible]	[illegible]	[illegible]	[illegible]	[illegible]	[illegible]	[illegible]	[illegible]	[illegible]	[illegible]	[illegible]	[illegible]
27 E5	[illegible]	[illegible]	[illegible]	[illegible]	[illegible]	[illegible]	[illegible]	[illegible]	[illegible]	[illegible]	[illegible]	[illegible]	[illegible]	[illegible]	[illegible]	[illegible]	[illegible]	[illegible]	[illegible]	[illegible]	[illegible]	[illegible]	[illegible]	[illegible]	[illegible]	[illegible]	[illegible]	[illegible]	[illegible]	[illegible]	[illegible]	[illegible]	[illegible]	[illegible]	[illegible]	[illegible]	[illegible]	[illegible]	[illegible]	[illegible]	[illegible]	[illegible]	[illegible]	[illegible]	[illegible]	[illegible]	[illegible]	[illegible]	[illegible]	[illegible]	[illegible]	[illegible]	[illegible]	[illegible]	[illegible]	[illegible]	[illegible]	[illegible]	[illegible]	[illegible]	[illegible]	[illegible]	[illegible]
28 90BRE	[illegible]	[illegible]	[illegible]	[illegible]	[illegible]	[illegible]	[illegible]	[illegible]	[illegible]	[illegible]	[illegible]	[illegible]	[illegible]	[illegible]	[illegible]	[illegible]	[illegible]	[illegible]	[illegible]	[illegible]	[illegible]	[illegible]	[illegible]	[illegible]	[illegible]	[illegible]	[illegible]	[illegible]	[illegible]	[illegible]	[illegible]	[illegible]	[illegible]	[illegible]	[illegible]	[illegible]	[illegible]	[illegible]	[illegible]	[illegible]	[illegible]	[illegible]	[illegible]	[illegible]	[illegible]	[illegible]	[illegible]	[illegible]	[illegible]	[illegible]	[illegible]	[illegible]	[illegible]	[illegible]	[illegible]	[illegible]	[illegible]	[illegible]	[illegible]	[illegible]	[illegible]	[illegible]	[illegible]
29 bcd	[illegible]	[illegible]	[illegible]	[illegible]	[illegible]	[illegible]	[illegible]	[illegible]	[illegible]	[illegible]	[illegible]	[illegible]	[illegible]	[illegible]	[illegible]	[illegible]	[illegible]	[illegible]	[illegible]	[illegible]	[illegible]	[illegible]	[illegible]	[illegible]	[illegible]	[illegible]	[illegible]	[illegible]	[illegible]	[illegible]	[illegible]	[illegible]	[illegible]	[illegible]	[illegible]	[illegible]	[illegible]	[illegible]	[illegible]	[illegible]	[illegible]	[illegible]	[illegible]	[illegible]	[illegible]	[illegible]	[illegible]	[illegible]	[illegible]	[illegible]	[illegible]	[illegible]	[illegible]	[illegible]	[illegible]	[illegible]	[illegible]	[illegible]	[illegible]	[illegible]	[illegible]	[illegible]	[illegible]
30 en	[illegible]	[illegible]	[illegible]	[illegible]	[illegible]	[illegible]	[illegible]	[illegible]	[illegible]	[illegible]	[illegible]	[illegible]	[illegible]	[illegible]	[illegible]	[illegible]	[illegible]	[illegible]	[illegible]	[illegible]	[illegible]	[illegible]	[illegible]	[illegible]	[illegible]	[illegible]	[illegible]	[illegible]	[illegible]	[illegible]	[illegible]	[illegible]	[illegible]	[illegible]	[illegible]	[illegible]	[illegible]	[illegible]	[illegible]	[illegible]	[illegible]	[illegible]	[illegible]	[illegible]	[illegible]	[illegible]	[illegible]	[illegible]	[illegible]	[illegible]	[illegible]	[illegible]	[illegible]	[illegible]	[illegible]	[illegible]	[illegible]	[illegible]	[illegible]	[illegible]	[illegible]	[illegible]	[illegible]
31 inv	[illegible]	[illegible]	[illegible]	[illegible]	[illegible]	[illegible]	[illegible]	[illegible]	[illegible]	[illegible]	[illegible]	[illegible]	[illegible]	[illegible]	[illegible]	[illegible]	[illegible]	[illegible]	[illegible]	[illegible]	[illegible]	[illegible]	[illegible]	[illegible]	[illegible]	[illegible]	[illegible]	[illegible]	[illegible]	[illegible]	[illegible]	[illegible]	[illegible]	[illegible]	[illegible]	[illegible]	[illegible]	[illegible]	[illegible]	[illegible]	[illegible]	[illegible]	[illegible]	[illegible]	[illegible]	[illegible]	[illegible]	[illegible]	[illegible]	[illegible]	[illegible]	[illegible]	[illegible]	[illegible]	[illegible]	[illegible]	[illegible]	[illegible]	[illegible]	[illegible]	[illegible]	[illegible]	[illegible]
32 eve	[illegible]	[illegible]	[illegible]	[illegible]	[illegible]	[illegible]	[illegible]	[illegible]	[illegible]	[illegible]	[illegible]	[illegible]	[illegible]	[illegible]	[illegible]	[illegible]	[illegible]	[illegible]	[illegible]	[illegible]	[illegible]	[illegible]	[illegible]	[illegible]	[illegible]	[illegible]	[illegible]	[illegible]	[illegible]	[illegible]	[illegible]	[illegible]	[illegible]	[illegible]	[illegible]	[illegible]	[illegible]	[illegible]	[illegible]	[illegible]	[illegible]	[illegible]	[illegible]	[illegible]	[illegible]	[illegible]	[illegible]	[illegible]	[illegible]	[illegible]	[illegible]	[illegible]	[illegible]	[illegible]	[illegible]	[illegible]	[illegible]	[illegible]	[illegible]	[illegible]	[illegible]	[illegible]	[illegible]
33 ro	[illegible]	[illegible]	[illegible]	[illegible]	[illegible]	[illegible]	[illegible]	[illegible]	[illegible]	[illegible]	[illegible]	[illegible]	[illegible]	[illegible]	[illegible]	[illegible]	[illegible]	[illegible]	[illegible]	[illegible]	[illegible]	[illegible]	[illegible]	[illegible]	[illegible]	[illegible]	[illegible]	[illegible]	[illegible]	[illegible]	[illegible]	[illegible]	[illegible]	[illegible]	[illegible]	[illegible]	[illegible]	[illegible]	[illegible]	[illegible]	[illegible]	[illegible]	[illegible]	[illegible]	[illegible]	[illegible]	[illegible]	[illegible]	[illegible]	[illegible]	[illegible]	[illegible]	[illegible]	[illegible]	[illegible]	[illegible]	[illegible]	[illegible]	[illegible]	[illegible]	[illegible]	[illegible]	[illegible]
34 cad	[illegible]	[illegible]	[illegible]	[illegible]	[illegible]	[illegible]	[illegible]	[illegible]	[illegible]	[illegible]	[illegible]	[illegible]	[illegible]	[illegible]	[illegible]	[illegible]	[illegible]	[illegible]	[illegible]	[illegible]	[illegible]	[illegible]	[illegible]	[illegible]	[illegible]	[illegible]	[illegible]	[illegible]	[illegible]	[illegible]	[illegible]	[illegible]	[illegible]	[illegible]	[illegible]	[illegible]	[illegible]	[illegible]	[illegible]	[illegible]	[illegible]	[illegible]	[illegible]	[illegible]	[illegible]	[illegible]	[illegible]	[illegible]	[illegible]	[illegible]	[illegible]	[illegible]	[illegible]	[illegible]	[illegible]	[illegible]	[illegible]	[illegible]	[illegible]	[illegible]	[illegible]	[illegible]	[illegible]
35 prd	[illegible]	[illegible]	[illegible]	[illegible]	[illegible]	[illegible]	[illegible]	[illegible]	[illegible]	[illegible]	[illegible]	[illegible]	[illegible]	[illegible]	[illegible]	[illegible]	[illegible]	[illegible]	[illegible]	[illegible]	[illegible]	[illegible]	[illegible]	[illegible]	[illegible]	[illegible]	[illegible]	[illegible]	[illegible]	[illegible]	[illegible]	[illegible]	[illegible]	[illegible]	[illegible]	[illegible]	[illegible]	[illegible]	[illegible]	[illegible]	[illegible]	[illegible]	[illegible]	[illegible]	[illegible]	[illegible]	[illegible]	[illegible]	[illegible]	[illegible]	[illegible]	[illegible]	[illegible]	[illegible]	[illegible]	[illegible]	[illegible]	[illegible]	[illegible]	[illegible]	[illegible]	[illegible]	[illegible]
36 BSH4	[illegible]	[illegible]	[illegible]	[illegible]	[illegible]	[illegible]	[illegible]	[illegible]	[illegible]	[illegible]	[illegible]	[illegible]	[illegible]	[illegible]	[illegible]	[illegible]	[illegible]	[illegible]	[illegible]	[illegible]	[illegible]	[illegible]	[illegible]	[illegible]	[illegible]	[illegible]	[illegible]	[illegible]	[illegible]	[illegible]	[illegible]	[illegible]	[illegible]	[illegible]	[illegible]	[illegible]	[illegible]	[illegible]	[illegible]	[illegible]	[illegible]	[illegible]	[illegible]	[illegible]	[illegible]	[illegible]	[illegible]	[illegible]	[illegible]	[illegible]	[illegible]	[illegible]	[illegible]	[illegible]	[illegible]	[illegible]	[illegible]	[illegible]	[illegible]	[illegible]	[illegible]	[illegible]	[illegible]
37 BSH9	[illegible]	[illegible]	[illegible]	[illegible]	[illegible]	[illegible]	[illegible]	[illegible]	[illegible]	[illegible]	[illegible]	[illegible]	[illegible]	[illegible]	[illegible]	[illegible]	[illegible]	[illegible]	[illegible]	[illegible]	[illegible]	[illegible]	[illegible]	[illegible]	[illegible]	[illegible]	[illegible]	[illegible]	[illegible]	[illegible]	[illegible]	[illegible]	[illegible]	[illegible]	[illegible]	[illegible]	[illegible]	[illegible]	[illegible]	[illegible]	[illegible]	[illegible]	[illegible]	[illegible]	[illegible]	[illegible]	[illegible]	[illegible]	[illegible]	[illegible]	[illegible]	[illegible]	[illegible]	[illegible]	[illegible]	[illegible]	[illegible]	[illegible]	[illegible]	[illegible]	[illegible]	[illegible]	[illegible]
38 al	[illegible]	[illegible]	[illegible]	[illegible]	[illegible]	[illegible]	[illegible]	[illegible]	[illegible]	[illegible]	[illegible]	[illegible]	[illegible]	[illegible]	[illegible]	[illegible]	[illegible]	[illegible]	[illegible]	[illegible]	[illegible]	[illegible]	[illegible]	[illegible]	[illegible]	[illegible]	[illegible]	[illegible]	[illegible]	[illegible]	[illegible]	[illegible]	[illegible]	[illegible]	[illegible]	[illegible]	[illegible]	[illegible]	[illegible]	[illegible]	[illegible]	[illegible]	[illegible]	[illegible]	[illegible]	[illegible]	[illegible]	[illegible]	[illegible]	[illegible]	[illegible]	[illegible]	[illegible]	[illegible]	[illegible]	[illegible]	[illegible]	[illegible]	[illegible]	[illegible]	[illegible]	[illegible]	[illegible]
39 Otp	[illegible]	[illegible]	[illegible]	[illegible]	[illegible]	[illegible]	[illegible]	[illegible]	[illegible]	[illegible]	[illegible]	[illegible]	[illegible]	[illegible]	[illegible]	[illegible]	[illegible]	[illegible]	[illegible]	[illegible]	[illegible]	[illegible]	[illegible]	[illegible]	[illegible]	[illegible]	[illegible]	[illegible]	[illegible]	[illegible]	[illegible]	[illegible]	[illegible]	[illegible]	[illegible]	[illegible]	[illegible]	[illegible]	[illegible]	[illegible]	[illegible]	[illegible]	[illegible]	[illegible]	[illegible]	[illegible]	[illegible]	[illegible]	[illegible]	[illegible]	[illegible]	[illegible]	[illegible]	[illegible]	[illegible]	[illegible]	[illegible]	[illegible]	[illegible]	[illegible]	[illegible]	[illegible]	[illegible]
40 w26	[illegible]	[illegible]	[illegible]	[illegible]	[illegible]	[illegible]	[illegible]	[illegible]	[illegible]	[illegible]	[illegible]	[illegible]	[illegible]	[illegible]	[illegible]	[illegible]	[illegible]	[illegible]	[illegible]	[illegible]	[illegible]	[illegible]	[illegible]	[illegible]	[illegible]	[illegible]	[illegible]	[illegible]	[illegible]	[illegible]	[illegible]	[illegible]	[illegible]	[illegible]	[illegible]	[illegible]	[illegible]	[illegible]	[illegible]	[illegible]	[illegible]	[illegible]	[illegible]	[illegible]	[illegible]	[illegible]	[illegible]	[illegible]	[illegible]	[illegible]	[illegible]	[illegible]	[illegible]	[illegible]	[illegible]	[illegible]	[illegible]	[illegible]	[illegible]	[illegible]	[illegible]	[illegible]	[illegible]
41 otd	[illegible]	[illegible]	[illegible]	[illegible]	[illegible]	[illegible]	[illegible]	[illegible]	[illegible]	[illegible]	[illegible]	[illegible]	[illegible]	[illegible]	[illegible]	[illegible]	[illegible]	[illegible]	[illegible]	[illegible]	[illegible]	[illegible]	[illegible]	[illegible]	[illegible]	[illegible]	[illegible]	[illegible]	[illegible]	[illegible]	[illegible]	[illegible]	[illegible]	[illegible]	[illegible]	[illegible]	[illegible]	[illegible]	[illegible]	[illegible]	[illegible]	[illegible]	[illegible]	[illegible]	[illegible]	[illegible]	[illegible]	[illegible]	[illegible]	[illegible]	[illegible]	[illegible]	[illegible]	[illegible]	[illegible]	[illegible]	[illegible]	[illegible]	[illegible]	[illegible]	[illegible]	[illegible]	[illegible]
42 57BAH	[illegible]	[illegible]	[illegible]	[illegible]	[illegible]	[illegible]	[illegible]	[illegible]	[illegible]	[illegible]	[illegible]	[illegible]	[illegible]	[illegible]	[illegible]	[illegible]	[illegible]	[illegible]	[illegible]	[illegible]	[illegible]	[illegible]	[illegible]	[illegible]	[illegible]	[illegible]	[illegible]	[illegible]	[illegible]	[illegible]	[illegible]	[illegible]	[illegible]	[illegible]	[illegible]	[illegible]	[illegible]	[illegible]	[illegible]	[illegible]	[illegible]	[illegible]	[illegible]	[illegible]	[illegible]	[illegible]	[illegible]	[illegible]	[illegible]	[illegible]	[illegible]	[illegible]	[illegible]	[illegible]	[illegible]	[illegible]	[illegible]	[illegible]	[illegible]	[illegible]	[illegible]	[illegible]	[illegible]
43 60HSN	[illegible]	[illegible]	[illegible]	[illegible]	[illegible]	[illegible]	[illegible]	[illegible]	[illegible]	[illegible]	[illegible]	[illegible]	[illegible]	[illegible]	[illegible]	[illegible]	[illegible]	[illegible]	[illegible]	[illegible]	[illegible]	[illegible]	[illegible]	[illegible]	[illegible]	[illegible]	[illegible]	[illegible]	[illegible]	[illegible]	[illegible]	[illegible]	[illegible]	[illegible]	[illegible]	[illegible]	[illegible]	[illegible]	[illegible]	[illegible]	[illegible]	[illegible]	[illegible]	[illegible]	[illegible]	[illegible]	[illegible]	[illegible]	[illegible]	[illegible]	[illegible]	[illegible]	[illegible]	[illegible]	[illegible]	[illegible]	[illegible]	[illegible]	[illegible]	[illegible]	[illegible]	[illegible]	[illegible]
44 cut	[illegible]	[illegible]	[illegible]	[illegible]	[illegible]	[illegible]	[illegible]	[illegible]	[illegible]	[illegible]	[illegible]	[illegible]	[illegible]	[illegible]	[illegible]	[illegible]	[illegible]	[illegible]	[illegible]	[illegible]	[illegible]	[illegible]	[illegible]	[illegible]	[illegible]	[illegible]	[illegible]	[illegible]	[illegible]	[illegible]	[illegible]	[illegible]	[illegible]	[illegible]	[illegible]	[illegible]	[illegible]	[illegible]	[illegible]	[illegible]	[illegible]	[illegible]	[illegible]	[illegible]	[illegible]	[illegible]	[illegible]	[illegible]	[illegible]	[illegible]	[illegible]	[illegible]	[illegible]	[illegible]	[illegible]	[illegible]	[illegible]	[illegible]	[illegible]	[illegible]	[illegible]	[illegible]	[illegible]
45 H2.0	[illegible]	[illegible]	[illegible]	[illegible]	[illegible]	[illegible]	[illegible]	[illegible]	[illegible]	[illegible]	[illegible]	[illegible]	[illegible]	[illegible]	[illegible]	[illegible]	[illegible]	[illegible]	[illegible]	[illegible]	[illegible]	[illegible]	[illegible]	[illegible]	[illegible]	[illegible]	[illegible]	[illegible]	[illegible]	[illegible]	[illegible]	[illegible]	[illegible]	[illegible]	[illegible]	[illegible]	[illegible]	[illegible]	[illegible]	[illegible]	[illegible]	[illegible]	[illegible]	[illegible]	[illegible]	[illegible]	[illegible]	[illegible]	[illegible]	[illegible]	[illegible]	[illegible]	[illegible]	[illegible]	[illegible]	[illegible]	[illegible]	[illegible]	[illegible]	[illegible]	[illegible]	[illegible]	[illegible]
46 ap	[illegible]	[illegible]	[illegible]	[illegible]	[illegible]	[illegible]	[illegible]	[illegible]	[illegible]	[illegible]	[illegible]	[illegible]	[illegible]	[illegible]	[illegible]	[illegible]	[illegible]	[illegible]	[illegible]	[illegible]	[illegible]	[illegible]	[illegible]	[illegible]	[illegible]	[illegible]	[illegible]	[illegible]	[illegible]	[illegible]	[illegible]	[illegible]	[illegible]	[illegible]	[illegible]	[illegible]	[illegible]	[illegible]	[illegible]	[illegible]	[illegible]	[illegible]	[illegible]	[illegible]	[illegible]	[illegible]	[illegible]	[illegible]	[illegible]	[illegible]	[illegible]	[illegible]	[illegible]	[illegible]	[illegible]	[illegible]	[illegible]	[illegible]	[illegible]	[illegible]	[illegible]	[illegible]	[illegible]
47 BK64	[illegible]	[illegible]	[illegible]	[illegible]	[illegible]	[illegible]	[illegible]	[illegible]	[illegible]	[illegible]	[illegible]	[illegible]	[illegible]	[illegible]	[illegible]	[illegible]	[illegible]	[illegible]	[illegible]	[illegible]	[illegible]	[illegible]	[illegible]	[illegible]	[illegible]	[illegible]	[illegible]	[illegible]	[illegible]	[illegible]	[illegible]	[illegible]	[illegible]	[illegible]	[illegible]	[illegible]	[illegible]	[illegible]	[illegible]	[illegible]	[illegible]	[illegible]	[illegible]	[illegible]	[illegible]	[illegible]	[illegible]	[illegible]	[illegible]	[illegible]	[illegible]	[illegible]	[illegible]	[illegible]	[illegible]	[illegible]	[illegible]	[illegible]	[illegible]	[illegible]	[illegible]	[illegible]	[illegible]
48 BK87	[illegible]	[illegible]	[illegible]	[illegible]	[illegible]	[illegible]	[illegible]	[illegible]	[illegible]	[illegible]	[illegible]	[illegible]	[illegible]	[illegible]	[illegible]	[illegible]	[illegible]	[illegible]	[illegible]	[illegible]	[illegible]	[illegible]	[illegible]	[illegible]	[illegible]	[illegible]	[illegible]	[illegible]	[illegible]	[illegible]	[illegible]	[illegible]	[illegible]	[illegible]	[illegible]	[illegible]	[illegible]	[illegible]	[illegible]	[illegible]	[illegible]	[illegible]	[illegible]	[illegible]	[illegible]	[illegible]	[illegible]	[illegible]	[illegible]	[illegible]	[illegible]	[illegible]	[illegible]	[illegible]	[illegible]	[illegible]	[illegible]	[illegible]	[illegible]	[illegible]	[illegible]	[illegible]	[illegible]
49 CF1-a	[illegible]	[illegible]	[illegible]	[illegible]	[illegible]	[illegible]	[illegible]	[illegible]	[illegible]	[illegible]	[illegible]	[illegible]	[illegible]	[illegible]	[illegible]	[illegible]	[illegible]	[illegible]	[illegible]	[illegible]	[illegible]	[illegible]	[illegible]	[illegible]	[illegible]	[illegible]	[illegible]	[illegible]	[illegible]	[illegible]	[illegible]	[illegible]	[illegible]	[illegible]	[illegible]	[illegible]	[illegible]	[illegible]	[illegible]	[illegible]	[illegible]	[illegible]	[illegible]	[illegible]	[illegible]	[illegible]	[illegible]	[illegible]	[illegible]	[illegible]	[illegible]	[illegible]	[illegible]	[illegible]	[illegible]	[illegible]	[illegible]	[illegible]	[illegible]	[illegible]	[illegible]	[illegible]	[illegible]
50 dPOU19	[illegible]	[illegible]	[illegible]	[illegible]	[illegible]	[illegible]	[illegible]	[illegible]	[illegible]	[illegible]	[illegible]	[illegible]	[illegible]	[illegible]	[illegible]	[illegible]	[illegible]	[illegible]	[illegible]	[illegible]	[illegible]	[illegible]	[illegible]	[illegible]	[illegible]	[illegible]	[illegible]	[illegible]	[illegible]	[illegible]	[illegible]	[illegible]	[illegible]	[illegible]	[illegible]	[illegible]	[illegible]	[illegible]	[illegible]	[illegible]	[illegible]	[illegible]	[illegible]	[illegible]	[illegible]	[illegible]	[illegible]	[illegible]	[illegible]	[illegible]	[illegible]	[illegible]	[illegible]	[illegible]	[illegible]	[illegible]	[illegible]	[illegible]	[illegible]	[illegible]	[illegible]	[illegible]	[illegible]
51 dPOU28	[illegible]	[illegible]	[illegible]	[illegible]	[illegible]	[illegible]	[illegible]	[illegible]	[illegible]	[illegible]	[illegible]	[illegible]	[illegible]	[illegible]	[illegible]	[illegible]	[illegible]	[illegible]	[illegible]	[illegible]	[illegible]	[illegible]	[illegible]	[illegible]	[illegible]	[illegible]	[illegible]	[illegible]	[illegible]	[illegible]	[illegible]	[illegible]	[illegible]	[illegible]	[illegible]	[illegible]	[illegible]	[illegible]	[illegible]	[illegible]	[illegible]	[illegible]	[illegible]	[illegible]	[illegible]	[illegible]	[illegible]	[illegible]	[illegible]	[illegible]	[illegible]	[illegible]	[illegible]	[illegible]	[illegible]	[illegible]	[illegible]	[illegible]	[illegible]	[illegible]	[illegible]	[illegible]	[illegible]
52 I-POU	[illegible]	[illegible]	[illegible]	[illegible]	[illegible]	[illegible]	[illegible]	[illegible]	[illegible]	[illegible]	[illegible]	[illegible]	[illegible]	[illegible]	[illegible]	[illegible]	[illegible]	[illegible]	[illegible]	[illegible]	[illegible]	[illegible]	[illegible]	[illegible]	[illegible]	[illegible]	[illegible]	[illegible]	[illegible]	[illegible]	[illegible]	[illegible]	[illegible]	[illegible]	[illegible]	[illegible]	[illegible]	[illegible]	[illegible]	[illegible]	[illegible]	[illegible]	[illegible]	[illegible]	[illegible]	[illegible]	[illegible]	[illegible]	[illegible]	[illegible]	[illegible]	[illegible]	[illegible]	[illegible]	[illegible]	[illegible]	[illegible]	[illegible]	[illegible]	[illegible]	[illegible]	[illegible]	[illegible]
53 exd	[illegible]	[illegible]	[illegible]	[illegible]	[illegible]	[illegible]	[illegible]	[illegible]	[illegible]	[illegible]	[illegible]	[illegible]	[illegible]	[illegible]	[illegible]	[illegible]	[illegible]	[illegible]	[illegible]	[illegible]	[illegible]	[illegible]	[illegible]	[illegible]	[illegible]	[illegible]	[illegible]	[illegible]	[illegible]	[illegible]	[illegible]	[illegible]	[illegible]	[illegible]	[illegible]	[illegible]	[illegible]	[illegible]	[illegible]	[illegible]	[illegible]	[illegible]	[illegible]	[illegible]	[illegible]	[illegible]	[illegible]	[illegible]	[illegible]	[illegible]	[illegible]	[illegible]	[illegible]	[illegible]	[illegible]	[illegible]	[illegible]	[illegible]	[illegible]	[illegible]	[illegible]	[illegible]	[illegible]
54 SPH-1	[illegible]	[illegible]	[illegible]	[illegible]	[illegible]	[illegible]	[illegible]	[illegible]	[illegible]	[illegible]	[illegible]	[illegible]	[illegible]	[illegible]	[illegible]	[illegible]	[illegible]	[illegible]	[illegible]	[illegible]	[illegible]	[illegible]	[illegible]	[illegible]	[illegible]	[illegible]	[illegible]	[illegible]	[illegible]	[illegible]	[illegible]	[illegible]	[illegible]	[illegible]	[illegible]	[illegible]	[illegible]	[illegible]	[illegible]	[illegible]	[illegible]	[illegible]	[illegible]	[illegible]	[illegible]	[illegible]	[illegible]	[illegible]	[illegible]	[illegible]	[illegible]	[illegible]	[illegible]	[illegible]	[illegible]	[illegible]	[illegible]	[illegible]	[illegible]	[illegible]	[illegible]	[illegible]	[illegible]
55 SPH-2A	[illegible]	[illegible]	[illegible]	[illegible]	[illegible]	[illegible]	[illegible]	[illegible]	[illegible]	[illegible]	[illegible]	[illegible]	[illegible]	[illegible]	[illegible]	[illegible]	[illegible]	[illegible]	[illegible]	[illegible]	[illegible]	[illegible]	[illegible]	[illegible]	[illegible]	[illegible]	[illegible]	[illegible]	[illegible]	[illegible]	[illegible]	[illegible]	[illegible]	[illegible]	[illegible]	[illegible]	[illegible]	[illegible]	[illegible]	[illegible]	[illegible]	[illegible]	[illegible]	[illegible]	[illegible]	[illegible]	[illegible]	[illegible]	[illegible]	[illegible]	[illegible]	[illegible]	[illegible]	[illegible]	[illegible]	[illegible]	[illegible]	[illegible]	[illegible]	[illegible]	[illegible]	[illegible]	[illegible]
56 SPH-2B	[illegible]	[illegible]	[illegible]	[illegible]	[illegible]	[illegible]	[illegible]	[illegible]	[illegible]	[illegible]	[illegible]	[illegible]	[illegible]	[illegible]	[illegible]	[illegible]	[illegible]	[illegible]	[illegible]	[illegible]	[illegible]	[illegible]	[illegible]	[illegible]	[illegible]	[illegible]	[illegible]	[illegible]	[illegible]	[illegible]	[illegible]	[illegible]	[illegible]	[illegible]	[illegible]	[illegible]	[illegible]	[illegible]	[illegible]	[illegible]	[illegible]	[illegible]	[illegible]	[illegible]	[illegible]	[illegible]	[illegible]	[illegible]	[illegible]	[illegible]	[illegible]	[illegible]	[illegible]	[illegible]	[illegible]	[illegible]	[illegible]	[illegible]	[illegible]	[illegible]	[illegible]	[illegible]	[illegible]
57 SPH-2C	[illegible]	[illegible]	[illegible]	[illegible]	[illegible]	[illegible]	[illegible]	[illegible]	[illegible]	[illegible]	[illegible]	[illegible]	[illegible]	[illegible]	[illegible]	[illegible]	[illegible]	[illegible]	[illegible]	[illegible]	[illegible]	[illegible]	[illegible]	[illegible]	[illegible]	[illegible]	[illegible]	[illegible]	[illegible]	[illegible]	[illegible]	[illegible]	[illegible]	[illegible]	[illegible]	[illegible]	[illegible]	[illegible]	[illegible]	[illegible]	[illegible]	[illegible]	[illegible]	[illegible]	[illegible]	[illegible]	[illegible]	[illegible]	[illegible]	[illegible]	[illegible]	[illegible]	[illegible]	[illegible]	[illegible]	[illegible]	[illegible]	[illegible]	[illegible]	[illegible]	[illegible]	[illegible]	[illegible]
58 So	[illegible]	[illegible]	[illegible]	[illegible]	[illegible]	[illegible]	[illegible]	[illegible]	[illegible]	[illegible]	[illegible]	[illegible]	[illegible]	[illegible]	[illegible]	[illegible]	[illegible]	[illegible]	[illegible]	[illegible]	[illegible]	[illegible]	[illegible]	[illegible]	[illegible]	[illegible]	[illegible]	[illegible]	[illegible]	[illegible]	[illegible]	[illegible]	[illegible]	[illegible]	[illegible]	[illegible]	[illegible]	[illegible]	[illegible]	[illegible]	[illegible]	[illegible]	[illegible]	[illegible]	[illegible]	[illegible]	[illegible]	[illegible]	[illegible]	[illegible]	[illegible]	[illegible]	[illegible]	[illegible]	[illegible]	[illegible]	[illegible]	[illegible]	[illegible]	[illegible]	[illegible]	[illegible]	[illegible]
59 Spou1	[illegible]	[illegible]	[illegible]	[illegible]	[illegible]	[illegible]	[illegible]	[illegible]	[illegible]	[illegible]	[illegible]	[illegible]	[illegible]	[illegible]	[illegible]	[illegible]	[illegible]	[illegible]	[illegible]	[illegible]	[illegible]	[illegible]	[illegible]	[illegible]	[illegible]	[illegible]	[illegible]	[illegible]	[illegible]	[illegible]	[illegible]	[illegible]	[illegible]	[illegible]	[illegible]	[illegible]	[illegible]	[illegible]	[illegible]	[illegible]	[illegible]	[illegible]	[illegible]	[illegible]	[illegible]	[illegible]	[illegible]	[illegible]	[illegible]	[illegible]	[illegible]	[illegible]	[illegible]	[illegible]	[illegible]	[illegible]	[illegible]	[illegible]	[illegible]	[illegible]	[illegible]	[illegible]	[illegible]
60 Spou2	[illegible]	[illegible]	[illegible]	[illegible]	[illegible]	[illegible]	[illegible]	[illegible]	[illegible]	[illegible]	[illegible]	[illegible]	[illegible]	[illegible]	[illegible]	[illegible]	[illegible]	[illegible]	[illegible]	[illegible]	[illegible]	[illegible]	[illegible]	[illegible]	[illegible]	[illegible]	[illegible]	[illegible]	[illegible]	[illegible]	[illegible]	[illegible]	[illegible]	[illegible]	[illegible]	[illegible]	[illegible]	[illegible]	[illegible]	[illegible]	[illegible]	[illegible]	[illegible]	[illegible]	[illegible]	[illegible]	[illegible]	[illegible]	[illegible]	[illegible]	[illegible]	[illegible]	[illegible]	[illegible]	[illegible]	[illegible]	[illegible]	[illegible]	[illegible]	[illegible]	[illegible]	[illegible]	[illegible]
61 Prox1	[illegible]	[illegible]	[illegible]	[illegible]	[illegible]	[illegible]	[illegible]	[illegible]	[illegible]	[illegible]	[illegible]	[illegible]	[illegible]	[illegible]	[illegible]	[illegible]	[illegible]	[illegible]	[illegible]	[illegible]	[illegible]	[illegible]	[illegible]	[illegible]	[illegible]	[illegible]	[illegible]	[illegible]	[illegible]	[illegible]	[illegible]	[illegible]	[illegible]	[illegible]	[illegible]	[illegible]	[illegible]	[illegible]	[illegible]	[illegible]	[illegible]	[illegible]	[illegible]	[illegible]	[illegible]	[illegible]	[illegible]	[illegible]	[illegible]	[illegible]	[illegible]	[illegible]	[illegible]	[illegible]	[illegible]	[illegible]	[illegible]	[illegible]	[illegible]	[illegible]	[illegible]	[illegible]	[illegible]
62 Prox2	[illegible]	[illegible]	[illegible]	[illegible]	[illegible]	[illegible]	[illegible]	[illegible]	[illegible]	[illegible]	[illegible]	[illegible]	[illegible]	[illegible]	[illegible]	[illegible]	[illegible]	[illegible]	[illegible]	[illegible]	[illegible]	[illegible]	[illegible]	[illegible]	[illegible]	[illegible]	[illegible]	[illegible]	[illegible]	[illegible]	[illegible]	[illegible]	[illegible]	[illegible]	[illegible]	[illegible]	[illegible]	[illegible]	[illegible]	[illegible]	[illegible]	[illegible]	[illegible]	[illegible]	[illegible]	[illegible]	[illegible]	[illegible]	[illegible]	[illegible]	[illegible]	[illegible]	[illegible]	[illegible]	[illegible]	[illegible]	[illegible]	[illegible]	[illegible]	[illegible]	[illegible]	[illegible]	[illegible]
63 Prox3	[illegible]	[illegible]	[illegible]	[illegible]	[illegible]	[illegible]	[illegible]	[illegible]	[illegible]	[illegible]	[illegible]	[illegible]	[illegible]	[illegible]	[illegible]	[illegible]	[illegible]	[illegible]	[illegible]	[illegible]	[illegible]	[illegible]	[illegible]	[illegible]	[illegible]	[illegible]	[illegible]	[illegible]	[illegible]	[illegible]	[illegible]	[illegible]	[illegible]	[illegible]	[illegible]	[illegible]	[illegible]	[illegible]	[illegible]	[illegible]	[illegible]	[illegible]	[illegible]	[illegible]	[illegible]	[illegible]	[illegible]	[illegible]	[illegible]	[illegible]	[illegible]	[illegible]	[illegible]	[illegible]	[illegible]	[illegible]	[illegible]	[illegible]	[illegible]	[illegible]	[illegible]	[illegible]	[illegible]

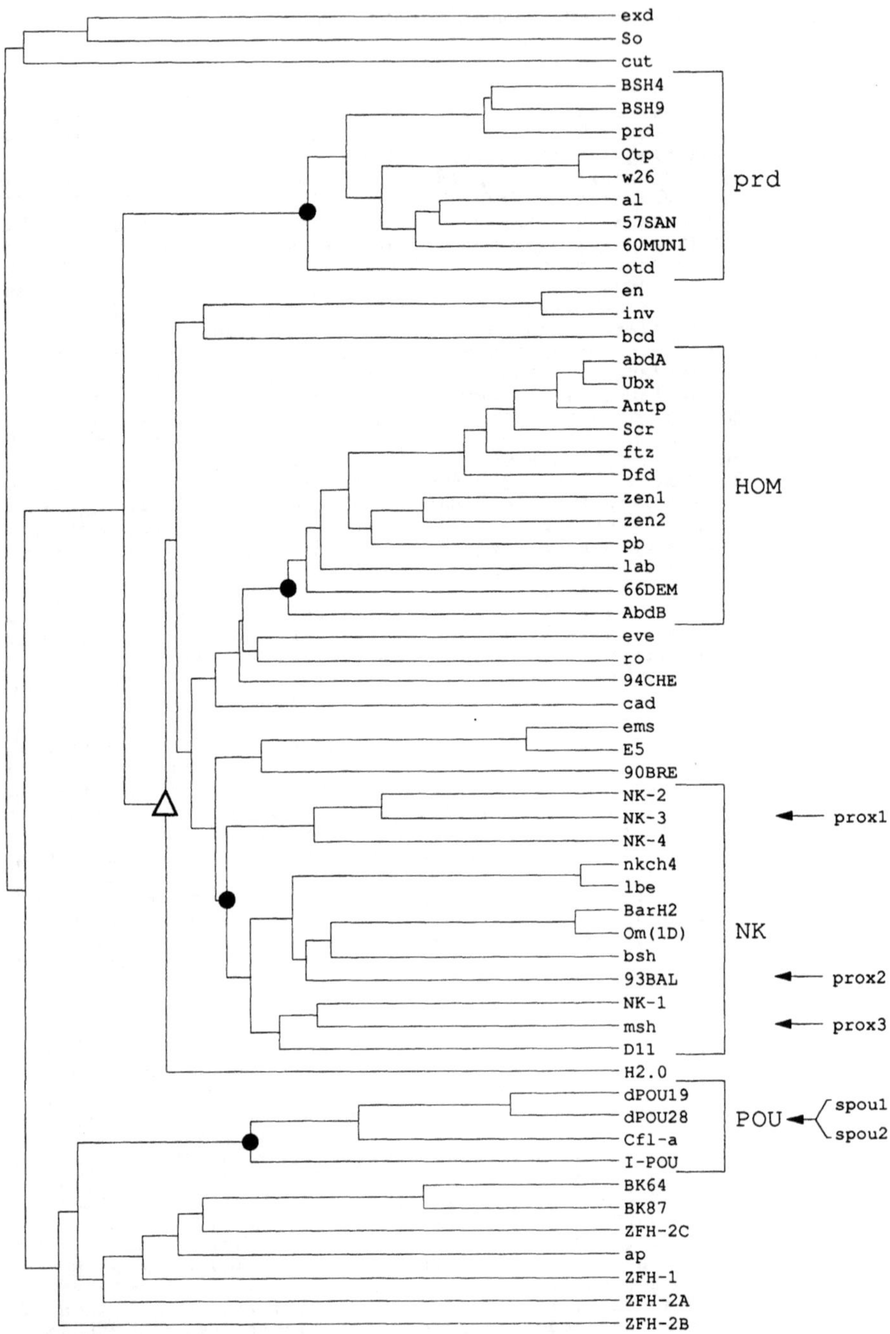

Fig. 2. A tree showing the classification of the homeobox genes of *Drosophila*. Based on the data summarized in Table 1, the comparative tree was generated by application of the PileUp system supplied with GCG software package. *Filled black circles* indicate the deviation points for the *prd* class, *HOM* class, *NK*-2 superclass, and *POU* class. Some of the genes included in the HOM class are not authentic homeotic genes. A *large triangle* indicates a large group whose members encode a hydrophobic amino acid at the 26th position. Other genes encode proline at the 26th position (for details, refer to the text). On the *right*, genes isolated from sponges are indicated at appropriate locations

number of homeobox genes have been shown to exist in mammalian species, homologous genes have proliferated by multiplication of preexistent genes during the evolution of vertebrates. Furthermore, in terms of the classification of homeobox genes, *Drosophila* seems to contain representatives of all known classes of homeobox genes, with the exception of a few that have evolved in higher animals. We compared the amino acid sequences of homeodomains encoded by homeobox genes of *Drosophila* and the calculated percentage of identity among the amino acid residues. Table 1 shows the results, from which we generated a comparative tree of *Drosophila* homeobox genes, as shown in Fig. 2. The combinations of the genes with more than 50% homology (in yellow) and with more than 60% homology (in orange and red) appear to be critical for classification of these genes. The HOX/HOM class forms the largest class, as shown in Fig. 2. If we define the *NK-2* superclass as indicated, it also has a large number of members. The percent of homology of amino acid residues among members of the *NK-2* superclass is relatively low. As we shall discuss below, however, the *NK-2* superclass might have preceded the HOM/HOX class during evolution. Moreover, the amino acid at the 50th position was clearly important for this classification. For example, many homeobox genes encode glutamine at this residue. When cysteine is the 50th residue, the gene should be classified into the *POU* class. The homeodomains in the *paired* class have serine as the 50th residue. Most of the strongly diverged homeobox genes, for example, *bcd, So, cut*, and *exd*, encode lysine, lysine, histidine, and glycine, respectively, at the 50th residue. It has been demonstrated that a change in the amino acid at the 50th residue alters the nucleotide sequence that is recognized by the homeodomain (Hanes and Brent 1989; Treisman et al. 1989).

The amino acid at the 26th position also appears to be important for our classification. In the case of the *Antennapedia* homeodomain, leucine at position 26 is involved in the formation of the hydrophobic core. The genes located in the central region of Fig. 2, from *en* to *H2.0*, have a hydrophobic amino acid, mainly leucine, at the 26th position. By contrast, all of the other genes encode proline at this position. Since the characteristics of a proline residue should be quite different from those of a leucine residue in terms of hydrophobicity and the effect on the structural configuration, strict conservation of either a hydrophobic amino acid or proline at the 26th position in these two major groups might be a good criterion for classification.

Many homeobox genes have been characterized by the presence of other functional motifs in addition to that of the homeodomain (Bürglin 1994). The HOM/HOX class encodes a hexapeptide, IYPWMK, upstream of the homeodomain that plays a role in interactions with other proteins (Phelan et al. 1995). The *paired* class of homeobox genes encodes a paired domain (Bopp et al. 1986). The *POU* class of homeobox genes encodes both a POU-specific domain and a POU-type homeodomain (Herr et al. 1988). As we shall discuss below, these functional motifs might have also been established at an early stage of animal evolution.

3
Homeobox Genes in Freshwater Sponges

3.1
Methods for the Identification of Homeobox Genes in Low Metazoans

Homeobox genes were initially isolated from *Drosophila* by analysis of mutants (Lewis 1978; Nüsslein-Volhard and Wieschaus 1980; Bender et al. 1983; Garber et al. 1983; Scott et al. 1983). Since the amino acid sequences of homeodomains are highly conserved in arthropods and vertebrates, many homeobox genes have been isolated by cross-hybridization with probes that correspond to genes from *Drosophila* (McGinnis et al. 1984b). Even within the *Drosophila* genome itself, more than 20 homeobox genes have been identified with a single probe by Southern hybridization under low-stringency conditions (Barad et al. 1988). However, the successful identification of homeobox genes by cross-hybridization with probes derived from *Drosophila* has been limited to Platyhelminthes and more complex members of the animal kingdom (Holland and Hogan 1986). Thus, two different methods have been developed for identification of homeobox genes in low metazoans.

Two hexapeptides, ELEKEF at the 15th through 20th positions and WFQNRR at the 48th through 53rd positions, are strongly conserved in the HOM/HOX class of homeobox genes, and degenerate oligonucleotides corresponding to these oligopeptides were synthesized as primers for the polymerase chain reaction (PCR; Murtha et al. 1991). PCR has been exploited for the identification of homeobox genes in many animals, including sponges (Murtha et al. 1991; Naito et al. 1993; Shenk et al. 1993; Seimiya et al. 1994).

Fig. 3A,B. Comparison of sequences between isolated clones and the primers used for PCR or the oligonucleotides used for screening. **A** Two kinds of mixed oligonucleotides were used for identification of homeobox genes by PCR. Both *prox1* and *prox2* were amplified with these primers. The nucleotides of the actual sequences that did not match those of the primers are *underlined*. *porx3* was not amplified by PCR. Another set of primers was used for amplification of POU-class homeobox genes. Both *spou-1* and *spou-2* were detected by use of these primers. Numbers of clones obtained are indicated. **B** A total of 28 clones was isolated by screening of a sponge genomic library that consisted of 300 000 independent clones (average size of inserts, 17 kb) with the HB-1 probe. Numerous *prox1*-, *prox2*-, and *prox3*-containing clones were obtained. Although other clones included sequences homologous to that of the probe, they appeared to be irrelevant clones. In the case of sph3, the sequences indicated were tandemly repeated ten times

```
A  PCR 5' primer for prox         PCR 3' primer for prox
                                                                Number of clones
      5'-GAATTAGAAAAAGAATT-3'        5'-  TCGTCGATTTTGAAACCA-3'     obtained      / number of clones
            GC G  G  G  G                 C TC TG  C  G                                analyzed

   prox1  GAGTTGGAACGGAGGTT       prox1   TCGTCTGTTCTGAAACCA          1

   prox2  GAATTGGAAAAAAAATT       prox2   TCTCCTATTTTGGAACCA          10

   prox3  GCACTCGAACGAAAGTT       prox3   TCGTCTATTCTGGAACCA          0 / 15


   PCR 5' primer for spou         PCR 3' primer for spou

      5'-ACAATATGTCGATTTGA-3'        5'-  TCTTCTATTACAAAACCA-3'
            T  T  CA G  C                 C GC GG  G  G
            G  C
            C

   spou-1 ACCATATGTAGGTTCGA       spou-1  TCTTCGGTTACAAAACCA          5

   spou-2 ACCATCTCTCGCTTTGA       spou-2  TCGCTTGTTGCAAAACCA          2 / 10


B  HB-1 probe                                5'-AAAATATGGTTTCAAAACAGAAG-3'
                                                 G  T    C  G  TC TC
                                                    G              G
                                                    C              C

   prox1                                        AAAATCTGGTTTCAGAACAGCCG         5

   prox2                                        AAAACATGGTTCCAAAATAGGAG         4

   prox3                                        AAAATTTGGTTCCAGAATAGACG         5 / 300 000

   sph3 (negative)                                  GCTTGGTTCCAGAAAGGCTAATGGGTATGAG   (x10)
                                                                 G     A
   sph6 (negative)              TTTGATCAGGATTAGTAGATTTGGATTCAGAACAGGTGCAATAGAAAGAAGGGAGGGGTTTTGTTGTAGGAC

   sph8 (negative) TAAGGAGTTCATTCACTTCTAGGCAGCCTGGAATGGTTTCAGAATAGAAGCATTTAACTC

   sph9 (negative) AACATTGTTAAAACAATGAACATTTCACCACACTGGTTTCAGAATAGGCTGGAAGAGTGCTTTTTATATAAAATA
```

Fig. 3A,B.

As long as two oligopeptides have been conserved, PCR can be successfully applied to the identification of target genes. We used this method to identify members of the POU class of homeobox genes in sponges, with POU-specific sequences as primers, as discussed below (Seimiya et al. 1997).

The degenerate oligonucleotide method, devised by Bürglin et al. (1989), has also proved to be very useful for the isolation of homeobox genes not only from animals but also from fungi and plants (Garcia-Fernandez et al. 1991; Ruberti et al. 1991; Oliver et al. 1992; Schena and Davis 1992; Schummer et al. 1992). This method assumes the presence of a highly conserved oligopeptide in the helix-3 region of the homeodomain, for example, KIWFQNRR, encoded by the HB-1 probe. PCR and the use of degenerate oligonucleotides both have advantages and disadvantages for the identification of homeobox genes. PCR allows the direct acquisition of sequence information in a short time, but amplification of many homeobox genes have proved impossible because of mismatching of sequences between primers and templates. The degenerate oligonucleotide method can reveal more homeobox genes than the PCR method, although some positive clones may be irrelevant. In Fig. 3, we summarize some examples of genes that were amplified by PCR and some sequences that were positive when the HB-1 probe was used to identify homeoboxes in the sponge genome.

The limitations of these two methods for the identification of homeobox genes were overcome recently by Balavoine (1996), who developed a ligation-mediated PCR technique. This method assumes only the presence of one stretch of conserved amino acids that is used for the design of the primer for one-directional PCR.

3.2 Homeobox Genes Isolated from Sponges

3.2.1 *prox1, prox2, and prox3*

We identified three homeobox genes in the freshwater sponge *Ephydatia fluviatilis* (Seimiya et al. 1994). Two of these genes, designated *prox1* and *prox2*, were identified by the PCR method. The third gene, designated *prox3*, was identified by the degenerate oligonucleotide method. For the analysis of the expression of these genes, gemmules of the freshwater sponge were allowed to hatch, and the cells were grown in a mineral-salts medium until a complete choanocyte chamber had formed. Seven days are required for the complete development of a sponge, and we arbitrarily divided this 7-day period into nine stages on the basis of visible changes in morphology, as follows: stage 1, before hatching; stage 2, immediately after hatching, i.e., 3 days after the start of incubation; and stage 9, complete formation of the choanocyte chamber 7 days after the start of incubation, with stages 3–8 representing sequential half-day intervals between stages 2 and 9. Transcripts of *prox1* and *prox2* were identified in cells at all stages of differentiation (Seimiya et al. 1994). This result suggests that *prox1* and *prox2* might encode general transcription factors that are expressed in all cells in the sponge.

3.2.2
spou-1 and spou-2

We also identified two homeobox genes that belong to the *POU* class, designated *spou-1* and *spou-2*, in the freshwater sponge (Seimiya et al. 1997). These genes each encode both a POU-specific domain and a POU-type homeodomain. By contrast to the results for *prox1* and *prox2*, transcripts of *spou-1* and *spou-2* were identified in cells only at specific stages during the differentiation of the sponge. *Spou-1* was expressed at stages 5 and 6, and *spou-2* was expressed at stages 5 through 8 (Seimiya et al. 1997). The stages at which the transcripts were identified appeared to correspond to the most active stages of development. For example, canals are formed at stages 5 and 6. The choanocyte, which is a flagellate cell crowned by a collar of cytoplasmic tentacles and is responsible for generating a water current, starts to appear at stages 6 and 7. The choanocyte chamber, which is a cavity enclosed by a group of choanocytes and oscula, is formed at stages 7 and 8. Since these specific cells and structures are retained in sponges at the adult stage, while the transcripts of *spou-1* and *spou-2* disappeared at stage 7 and at stage 9, respectively, the products of *spou-1* and *spou-2* might perhaps function as transcription factors that play roles in the determination of cell fate during differentiation, rather than as housekeeping proteins.

3.2.3
Other Homeobox Genes

The presence of homeobox genes in sponges has also been reported by other investigators. Degnan et al. (1995) identified two homeobox genes, designated *SpoxH1* and *SpoxH2*, from the sponge *Hariclona*, as well as another homeobox gene, desiganated *SpoxTA1*, from the sponge *Tethya aurentia*. Only partial sequences of the regions sandwiched by the primers used for PCR were reported. Kruse et al. (1994) described the sequence of a homeoboxlike gene, designated *SHOX*, in the marine sponge *Geodia cydonium*. However, it seems unlikely that *SHOX* is a homeobox gene since the critical sequence *WFQNRR*, at positions 48 through 53 in standard homeodomains, is replaced by the irrelevant sequence VRQLRA in the protein encoded by *SHOX*.

4
Comparison of the Homeobox Genes of Sponges with Those of Other Animals

4.1
Comparison of the Amino Acid Sequences of the Homeodomains Encoded by *prox1, prox2* and *prox3* of Sponges with Those of Other Animals

We compared the amino acid sequences of the homeodomains encoded by *proxl, prox2* and *prox3* with those of all the published sequences of homeodomains from various kinds of animals. As shown in Fig 4, the amino acid sequence of the homeodomain encoded by *NK-3 (bap)* of *Drosophila* showed the highest degree of homology in terms of amino acid residues to the homeodomain encoded by *proxl*. Sequences encoded by *NK-3*-related genes, such as *NK-4 (tin)* and *NK-2 (vnd)* of *Drosophila*, and *Dth2* of planaria, also showed high degrees of homology that ranged between 63–65%.

In the case of *prox2, 93BAL (HOX11)* of *Drosophila* showed the highest degree of homology. *Om(1D), BarH2*, and *bsh* of *Drosophila* showed 62% homology to *prox2* at the amino acid level. The amino acid sequence of the *prox3* homeodomain was 77%, 73%, and 65% identical to that of the *msh* gene of human, *Drosophila*, and hydra, respectively.

Similarities between the homeodomains encoded by sponge genes and those encoded by genes of other animals can be found not only in the high degree of homology among overall sequences but also at specific positions that appear to be critical for the recognition of specific nucleotide sequences. For example, threonine at the 46th position appears to be rare among the homeodomains that have been characterized to date, but *prox2, 93BAL, bsh, BarH2*, and *Om(1D)* all encode threonine at this position. In the case of the *msh* gene, not only the sequence of the homeodomain but also its short flanking sequence is conserved. CXLRKHKX and EAE (I/L) E are the conserved sequences in the amino-terminal and carboxy-terminal flanking regions, respectively (Schummer et al. 1992). *prox3* encodes STINKQKK and EAEAE at these positions, respectively (Seimiya et al. 1994). Thus, it appears that the *proxl, prox2*, and *prox3* genes identified in the sponge can be assigned to classes of homeodomains found in higher organisms. Therefore, several distinct classes of homeobox genes previously identified in higher organisms seem already to have become established when metazoans appeared during evolution. It is of considerable interest in this context that the classes to which *proxl, prox2*, and *prox3* can be assigned appear to form a large superclass, namely, NK-2.

When we designed our primers, we assumed that homeobox genes that encode the sequence ELEKEF might be present in sponges. However, *proxl* encoded ELERRF (four nucleotide mismatches/17 nucleotides), and *prox2* encoded ELEKKF (one uncleotide mismatch/17 nucleotides) in the correspond-

		1	2	3
1 prox1	KRRPRALFSH AQVYELERRF AVQKYLTAHE QSKLATVLHL TETQVKIWFQ NRRYKSKRQQ			
2 NK-3 (bap)	-K-S--A--- ---F------ -Q-R--SGP- R-EM-KS-R- ---------- -----T--K-	72		(%)
3 NK-4 (tin)	--K--V---Q ---L---C-- RLK----GA- REII-QK-N- SA-------- --------GD	65	58	
4 NK-2 (vnd)	--KR-V--TK --T------- RQ-R--S-P- REH--SLIR- -P-------- -H---T--A-	63	67	60

		1	2	3	4
1 prox2	RKKARTAFSR EQVAELEKKF QNKKYLSSTE RGELAEKLKL SDMQVKTWFQ NRRMKYKRQS				
2 93BAL	---P--S-T- I-------R- HKQ---A-A- -AA--RG--M T-A------- ---T-WR--T	65			(%)
3 bsh	-R----V--D P-LSG---R- EGQR---PP- -V---TA-G- -ET------- -----H-K-L	62	52		
4 BarH2	QR------TD H-LQT---S- ERQ----VQD -M---N--E- --C-----Y- ---T-E---T	62	57	62	
5 Om (1D)	QR------TD H-LQT---S- ERQ----VQ- -Q---H--D- --C-----Y- ---T-WM--T	62	58	62	92

		1	2	3
1 prox3	DRKPRTPFTS TQLIALERKF RQQKYLSVAE RAEFAEYLKL TETQVKIWFQ NRRAKEKRLH			
2 msh (human)	N--------T A--LR----- --KQ---I-- ----SSS-S- ---------- -----A---Q	77		(%)
3 msh (Drosophila)	N--------T Q--LS--K-- -EKQ---I-- ----SSS-R- ---------- -----A---Q	73	92	
4 msh (hydra)	N-------SV N--LT--Q-- KRKQ---IS- ---LS-L-R- ----I----- -----Q--SK	65	75	77

Fig. 4. Comparison of the amino acid sequences of homeodomains encoded by *prox1*, *prox2*, and *prox3* with those of various other homeodomains. *Dashes* indicate the same amino acids as those in the respective sponge homeodomain. References for these data can be found elsewhere (Seimiya et al. 1994)

ing regions, even though the mismatched oligonucleotides functioned as primers. Among the products of PCR, ten clones corresponded to *prox2*, and one clone corresponded to *prox1*. When we used the degenerate oligonucleotide method, we isolated numerous *prox1-*, *prox2-* and *prox-3*-containing clones from the sponge genomic library that we had prepared. Since the methods that we used for the isolation of homeobox genes from sponges had various limitations, as discussed above, it is possible that many homeobox genes remain to be identified in the sponge genome. However, the methods that involved PCR and degenerate oligonucleotides were based on the presumed presence of the sequences ELEKEF, WFQNRR, and KIWFQNRR, which are strongly conserved in the products of homeobox genes that belong to the HOX/HOM class. Therefore, it seems unlikely that the HOM/HOX class of homeobox genes existed when the first metazoans appeared in the animal kingdom. Conversely, it seems likely that the *NK-2* superclass of homeobox genes preceded the HOM/HOX class during evolution.

4.2
Comparison of Amino Acid Sequences of POU-Specific Domains and POU-Type Homeodomains Encoded by *spou-1* and *spou-2* of Sponge with Those of Other Animals

We compared the amino acid sequences of the POU-specific domain and the POU-type homeodomain encoded by *spou-1* and *spou-2* with those encoded by the homeobox genes in the POU class whose sequences are available, as shown in Fig. 5. The percentage of similarity among amino acids is also indicated. The *spou-1* gene appears to be closest to the human *Pit-1* gene (*hPit-I*) among all the homeobox genes that are analyzed. The amino acid sequences of the POU-specific domain and the POU-type homeodomain encoded by *spou-1* were found to be 76 and 67% homologous to those encoded by *hPit-1*, respectively. The homeobox genes in the POU class have been further classified into six families, designated I to VI (Rosenfeld 1991; Bürglin 1994). The *hPit-I* gene represents family I. The degree of similarity between the amino acid sequences encoded by *spou-1* and *hPit-1* appears, in general, to be higher than those between sequences encoded by the genes in other families. We observed the similarity between *spou-1* and *hPit-1* not only at the amino acid level but also at those specific residues at which unique amino acids have been found in *hPit-1* exclusively among all the POU-class homeobox genes. For example, we found glutamic acid at the 5th, threonine at the 28th, asparagine at the 29th, valine at the 37th, histidine at the 38th, asparagine at the 52nd, and alanine at the 60th position in the POU-specific domains encoded by both *spou-1* and *hPit-1*. Similarity between *spou-1* and *hPit-1* was also found in the length of the linker that connects the POU-specific domain with the POU-type homeodomain. Although the length of the linker varies among members of the POU class of homeobox genes, the genes that belong to the same family encode linkers with the same respective numbers of amino acid residues, with the exception of

POU specific domain

	Sequence	1	2	3	4	5	6	7	8	9	10
Consensus	EDTPDSRELEQFAKQFKQRRIKLGFTQADVGLALGTLYGNVFGSLSQTTICRFEALQLSFKNMCKLKPLLNKWLEEAE										
1 spou-1	P-SDEVKD--S--AS--S--M---Y--TN--Q--REVH-TD- ---------N----Y--AQ----I-E-------										
2 spou-2	-Y--AFV--K---EE--T---R--Y--GA--QS-ADRGFSN- A-S--S---Q----PT-AAAI-QI-E---L---	54									
3 hPit-1 (I)	M-S-EI----K--NE--V------Y--TN--E--AAVH-SE- ---------N-------A----AI-S-------	76	59								
4 hOct-2 (II)	-EPS-LE-----RT------------G-----M-K----D- -----S-----N-------------E---ND--	55	49	56							
5 BmpouM1 (III)	----T-DD--A------------------------------- -----------------------Q------D	60	47	63	74						
6 DjPOU1 (III)	--L-S-DD------M---------Y----------------- ---------------------R---Q---H--D	59	47	62	73	91					
7 unc-86 (IV)	DMDT-P-Q--T--EH---------V------K--AH-KMPGV-----S------S-T--HN--VA---I-HS---K--	50	46	53	53	55	53				
8 tI-POU (IV)	DTDT-P----A--ER---------V------K--AN-KLPGV-AV--S------S-T--HN--IA---I-QA------	51	49	55	54	59	55	85			
9 xOct-60 (V)	--GMTLE-M-E---EL--K-VA--Y--G-I-H---I---KM- ---------S---T-----------EQ--G---	55	46	58	63	64	63	46	49		
10 mOct-3/4 (V)	DMKALQK-------LL--K--T--Y-------T--V-F-K-- --------------L------R---E--V---D	58	47	58	60	68	71	50	51	64	
11 ceh-18 (VI)	D-RI-MN---A--QT--KQ---F----G---V---KR--TD- -----S-----N---------R---KE--ADV-	51	41	53	74	65	64	47	50	54	54

(%)

POU homeodomain

	Sequence	1	2	3	4	5	6	7	8	9	10
Consensus	KRKKRTSIEVNVKGALEKHFLKCPKPSSQEITSIADSLQLEKEVVRVWFCNRRQKEKRMT										
1 spou-1	R--R----G-GAQET--R--INQ-----SD-NK-----G---------------R---VP										
2 spou-2	-----AVFTSQT-AL--N--AVN-RLNRAT-EMLSGK-D-LP-EI-------K---A-QSG	35									
3 hPit-1 (I)	---R--T-SIAA-D---R--GEQN-------MRM-EE-N---------------R---VK	67	37								
4 hOct-2 (II)	R--------T--RF----S--ANQ--T-E--LL--EQ-HM----I--------------IN	55	37	55							
5 BmpouM1 (III)	----------S------Q--H-Q----A-----L-------------------------	65	43	65	60						
6 DjPOU1 (III)	----------A---SI--SS-M-LS---A-D-S-L-EK-S------------------I-	57	42	60	63	75					
7 unc-86 (IV)	-KR-----AAPE-RE--QF-KQQ-R--GER-A----R-D-K-N---------Q---Q--DF	50	38	47	45	58	53				
8 tI-POU (IV)	-KR-----AAPE-RS--AY-AVQ-R--GEK-AA--EK-D-K-N---------Q---Q--IV	48	43	48	48	53	57	82			
9 xOct-60 (V)	---M--CFDTVL--Q--G--MCNQ--GAR-L-E--KE-S---D--------------SKF	48	38	53	52	58	55	43	42		
10 mOct-3/4 (V)	RKR------NR-RWS--TM---------L-Q--H--NQ-G---D-------------G--SS	55	33	48	58	62	58	53	47	47	
11 ceh-18 (VI)	R-R---NLDM-QRN--DTF-ALN-R-DHDKM-D--N--E-DRD-------------MR-VD	43	33	38	45	45	40	43	45	43	52

(%)

Fig. 5. Comparison of the amino acid sequences of POU-specific domains and POU homeodomains of *spou-1* and *spou-2* with those of various other POU-class genes. References for these data can be found elsewhere (Seimiya et al. 1997)

genes in family V (Bürglin 1994). The amino acid sequence of the POU-specific domain encoded by *spou-2* was 59% homologous to that encoded by *hPit-1*. This value might seemed rather low, but it was still the highest among the values obtained with all other subclasses.

Although the extent of homology among nucleotide sequences in the regions that we used as probes for *spou-1* and *spou-2* was relatively low, 53.7% in 294 bp, the probes did cross-hybridize with each other in Southern hybridization of sponge genomic DNA under low-stringency conditions. Since, under these conditions, the probes detected no more bands or only one more very weak band, the total number of homeobox genes in sponges that belong to the POU class, even if more genes exist, seems to be low. Moreover, even if we did overlook other POU genes, they might have very divergent sequences. The homeobox genes in the POU class that have been found in contemporary metazoans should be descendants of a single ancient gene. Thus, we propose that family I might represent the most ancient subclass among the six families of homeobox genes in this class. Prior to our studies, the planarian was the lowest animal in which POU-class homeobox genes has been isolated (Orii et al. 1993). Tow of these genes isolated from the planarian have been classified as members of family III. Although, to date, POU-class homeobox genes belonging to family I have been identified in vertebrates (Yamada et al. 1993) and sponges (Seimiya et al. 1997), it is possible that they exist in all metazoans.

Our results clearly show that, in addition to homeodomains, POU-specific domains were also established at an early stage of animal evolution, as might also be true for other functional motifs that have been found in homeobox genes. In fact, a gene that encodes a paired domain has also been identified in sponges (W.J. Gehring pers. comm.).

5 Principles Associated with the Conservation and Diversification of Homeobox Genes During Animal Evolution

Our results indicate clearly that, when metazoans appeared during the course of evolution, the multiple and distinct classes of homeobox genes that have been identified in higher organisms already existed. It remains, however, to be determined whether the homeobox genes that have been identified in sponges really represent ancient established homeobox genes. In other words, were the *NK-3*, *93BAL*, and *msh* classes of homeobox genes established earliest among the various classes of genes that belong to the NK-2 superclass? Was the NK-2 superclass established before the HOM/HOX class? Did family I of POU-class homeobox genes appear earliest among the six families? If such was the case, have these classes of homeobox genes been retained in all extant metazoans? We propose that the following hypothesis deserves strong consideration: all genes that exist in contemporary organisms are descendants of sets of putative primordial genes that were present in ancient organisms. All of these extant genes were created by multiplication and/or exon shuffling of primordial genes, with subsequent diversification by mutation and selection during evolution (Kurosawa and Hashimoto 1996). The number of primordial genes from which all contemporary genes are descended has been estimated to be less than one thousand (Chothia 1994). Therefore, one issue that is critical in attempts to answer the above questions is whether the similarities between the genes from sponges and *Drosophila* are high enough to allow us to consider the genes to be homologous to one another and to judge how frequently gene loss occurred during evolution.

Increases in the number of genes should have provided one of the major driving forces in the creation of more complex organisms during evolution. It has been argued that developmentally relevant genes have become integrated into developmental pathways that are hierarchical and highly interdependent and, thus, such genes cannot readily mutate or assume new functions without disruption of the overall developmental plan. Gene duplication provides a way around this impasse. It allows old developmental relationships to be maintained while at the same time newly duplicated genes are incorporated into new pathways and relationships (Ohno 1970; Ruddle et al. 1994). Our results seemed to support such an argument. During evolution, homeobox genes and POU-class homeobox genes multiplied, mutations were introduced, and selection operated to ensure either conservation or diversification of the respective genes. Mutations might have been introduced at a rather constant frequency. Therefore, whatever differences we observe might have been maintained as a consequence of selection. Figure 6 shows a model of two hypothetical modes of diversification of multiplied genes. In this model, after the evolutionary separation of animals into different species, the duplication of a specific gene occurred. In cases of type I, the sequence of one gene was conserved, and that of the other gene diverged from the sequence of the original gene. In cases of type II, the sequences of both genes diverged equally from that of the original gene. The results for the homeobox genes of sponges appear to suggest type I as the mode for the evolution of homeobox genes. The explanation for this phenomenon during evolution might reflect the roles of homeobox genes: they appear to be involved in the regulation of expression of essential genes in each animal via recognition of specific DNA sequences, as well as via protein-protein interactions. Therefore, at least one gene of a particular type should retain the characteristics of the prototype gene after its multiplication during evolution, with the other sibling and descendant genes gaining the freedom to mutate. Manak and Scott (1994) proposed that the characteristics of the various classes of homeobox genes, once established, have been strongly conserved both structurally and functionally. Plausible examples of gene loss during evolution have been clearly implicated at the loci of Hox genes. These loci in higher vertebrates encode four times more homeobox genes than those encoded in amphioxus, which has been proposed as an ancestor of vertebrates (Garcia-Fernandez and Holland 1994; Ruddle et al. 1994). It is clear that some of the quadruplicated genes have been lost during vertebrate evolution. In the

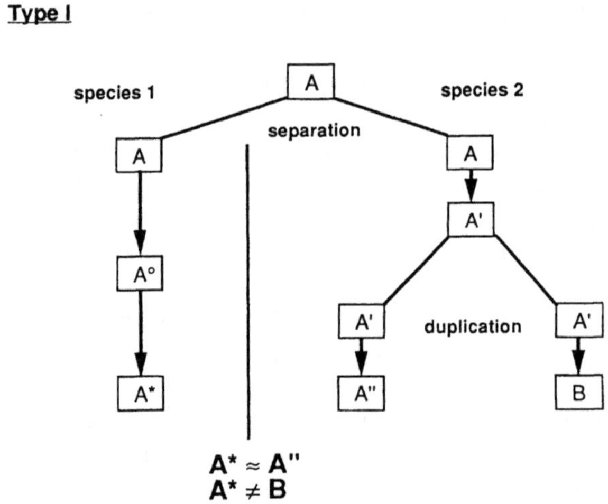

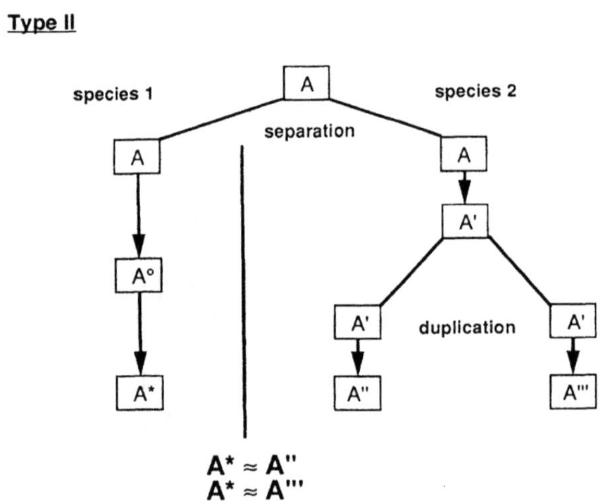

Fig. 6. Two representative models for diversification via duplicated genes. After separation of species, the primordial gene is duplicated in *species 2*. In cases designated *type I*, the sequence of one gene is conserved while that of the other gene changes rapidly. In cases designated *type II*, the sequences of both genes change at an equal rate. If we assume that the sequence of this gene is conserved in *species 1*, a comparison of the sequences of the genes between species 1 and species 2 should reveal whether the mode of diversification was type I or type II

case of such gene loss, however, more than one gene that represented a set of prototype genes was always retained after the gene loss. Thus, even if we have overlooked many homeobox genes in sponges and, furthermore, even if many genes that were established at an early stage of animal evolution have been lost in contemporary animals as a consequence of their evolution, we can conclude that the classes of homeobox genes found in sponges, such as *NK-3, msh,* and family I of the POU class, do represent ancient established genes in metazoans.

6 Conclusion

The homeodomains encoded by homeoboxes are one of the major types of structural motif that can bind to DNA. Homeobox-containing genes have been identified not only in animals but also in fungi and plants. Similarities among the amino acid sequences of homeodomains have allowed us to divide the homeobox genes in higher animals into many classes and several superclasses. None of the homeobox genes in fungi and plants exhibit a high degree of homology to any class of homeobox genes from animals.

We isolated five kinds of homeobox genes from the freshwater sponge *E. fluviatilis.* Three of these genes, *prox1, prox2,* and *prox3,* exhibited strong homology to the *NK3, 93BAL,* and *msh* genes, respectively, of *Drosophila.* These three genes belong to the NK-2 superclass. Since transcripts of *prox1* and *prox2* were identified at all stages of differentiation of the sponge, it seems possible that they might encode general transcription factors that are expressed in all cells in the sponge. Two POU-class homeobox genes, *spou-1* and *spou-2,* were also isolated from the sponge. Each of these genes encoded a POU-specific domain and a POU-class homeodomain. By contrast to the results for *prox1* and *prox2,* transcripts of the *spou-1* and *spou-2* genes were identified in cells only at specific stages during the differentiation of the sponge. Thus, they might encode transcription factors that play roles in the determination of cell fate during differentiation. The POU-type homeobox genes have been classified into six families, and the *spou-1* gene was strongly homologous to the human *Pit-1* gene, which has been classified as a member of family I. The *spou-2* gene also exhibited the greatest similarity to human *Pit-1,* but the extent of similarity was rather limited. These results clearly indicate that the multiple and distinct classes of homeobox genes that have been identified in higher organisms already existed when the metazoans appeared during the course of evolution. The characteristics of each class of homeodomains have been retained during animal evolution. From these observations, we have proposed some general principles related to the conservation and diversification of homeobox genes during evolution. The strong conservation of homeobox genes in all metazoans examined to date also implies the monophyly of metazoans, including sponges.

Acknowledgments. The authors thank Ms. A. Suzuoki for preparation of the manuscript. This work was supported in part by a Grant-in-Aid for Scientific Research on Priority Areas, Ministry of Education, Science and Culture, Japan; Fujita Health University.

References

Akam M (1989) Hox and Hom: homologous gene clusters in insects and vertebrates. Cell 57: 347–349

Bachiller DA, Macias D, Duboule D, Morata G (1994) Conservation of a functional hierarchy between mammalian and insect Hox/HOM genes. EMBO J 13: 1930–1941

Balavoine G (1996) Identification of members of several homeobox genes in a planarian using a ligation-mediated polymerase chain reaction technique. Nucleic Acids Res 24: 1547–1553

Barad M, Jack T, Chadwick R, McGinnis W (1988) A novel, tissue-specific, *Drosophila* homeobox gene. EMBO J 7: 2151–2161

Bender W, Akam M, Karch F, Beachy PA, Peifer M, Spierer P, Lewis EB, Hogness DS (1983) Molecular genetics of the bithorax complex in *Drosophila melanogaster*. Science 221: 23–29

Billeter M, Qian YQ, Otting G, Müller M, Gehring W, Wüthrich K (1993) Determination of the nuclear magnetic resonance solution structure of an Antennapedia homeodomain-DNA complex. J Mol Biol 234: 1084–1097

Bopp D, Burri M, Baumgartner S, Frigerio G, Noll M (1986) Conservation of a large protein domain in the segmentation gene *paired* and in functionally related genes of *Drosophila*. Cell 47: 1033–1040

Bürglin TR (1994) A comprehensive classification of homeobox genes. In: Douboule D (ed) Guidebook to the homeobox genes. A Sambrook & Tooze Publication at Oxford University Press, Oxford, pp 27–71

Bürglin TR, Finney M, Coulson A, Ruvkun G (1989) *Caenorhabditis elegans* has scores of homeobox-containing genes. Nature 341: 239–243

Cestka TA, Lamers M, Monaci P, Nicosia A, Cortese R, Suck D (1993) The X-ray structure of an atypical homeodomain present in the rat liver transcription factor LFB1/HNF1 and implifications for DNA binding. EMBO J 12: 1805–1810

Chothia C (1994) Protein families in the metazoan genome. Development (Suppl): 27–33

Christen R, Ratto A, Baroin A, Perasso R, Grell KG, Adoutte A (1991) An analysis of the origin of metazoans, using comparisons of partial sequences of the 28S RNA, reveals an early emergence of triploblasts. EMBO J 10: 499–503

Degnan BM, Degnan SM, Giusti A, Morse DE (1995) A hox/hom homeobox gene in sponges. Gene 155: 175–177

Dessain S, McGinnis W (1993) *Drosophila* homeobox genes. In: Wassarman PM (ed) Adv Dev Biochem 2: 1–55

Field KG, Olsen GJ, Lane DJ, Giovannoni SJ, Ghiselin MT, Raff EC, Pace NR, Raff RA (1988) Molecular phylogeny of the animal kingdom. Science 239: 748–753

Frasch M, Chen X, Lufkin T (1995) Evolutionary-conserved enhancers direct region-specific expression of the murine Hoxa-1 and Hoxa-2 loci in both mice and *Drosophila*. Development 121: 957–974

Garber RL, Kuroiwa A, Gehring WJ (1983) Genomic and cDNA clones of the homeotic locus Antennapedia in *Drosophila*. EMBO J 2: 2027–2036

Garcia-Fernandez J, Holland PWH (1994) Archetypal organization of the amphioxus Hox gene cluster. Nature 370: 563–566

Garcia-Fernandez J, Baguna J, Salo E (1991) Planarian homeobox genes: cloning, sequence analysis, and expression. Proc Natl Acad Sci USA 88: 7338–7342

Hanes SD, Brent R (1989) DNA specificity of the bicoid activator protein is determined by homeodomain recognition helix residue 9. Cell 57: 1275–1283

Hayashi S, Scott MP (1990) What determines the specificity of action of *Drosophila* homeodomain proteins? Cell 63: 883–894

Herr W, Sturm RA, Clerc RG, Corcoran LM, Baltimore D, Sharp PA, Ingraham HA, Rosenfeld MG, Finney M, Ruvkun G, Horvitz HR (1988) The POU domain: a large conserved region in the mammalian *pit-1*, *oct-1*, *oct-2*, and *Caenorhabditis elegans unc-86* gene products. Genes Dev 2: 1513–1516

Hirsch JA, Aggarwal AK (1995) Structure of the Even-skipped homeodomain complexed to AT-rich DNA: new perspectives on homeodomain specificity. EMBO J 14: 6280–6291

Holland PWH, Hogan BLM (1986) Phylogenetic distribution of Antennapedia-like homeo boxes. Nature 321: 251–253

Hyman LH (1951) The invertebrates: Plathyhelminthes and Rhynchocoela, vol 2. McGraw-Hill, New York

Jagla K, Stanceva I, Dretzen G, Bellard F, Bellard M (1994) A distinct class of homeodomain proteins is encoded by two sequentially expressed *Drosophila* genes from the 93D/E cluster. Nucleic Acids Res 22: 1202–1207

Kappen C, Schughart K, Ruddle FH (1993) Early evolutionary origin of major homeodomain sequence classes. Genomics 18: 54–70

Kenyon C, Wang B (1991) A cluster of Antennapedia-class homeobox genes in a nonsegmanted animal. Science 253: 516–517

Kissinger CR, Liu B, Martin-Blanco E, Kornberg TB, Pabo CO (1990) Crystal structure of an engrailed homeodomain-DNA coplex at 2.8 Å resolution: a framework for understanding homeodomain-DNA interactions. Cell 63: 579–590

Klemm JD, Rould MA, Aurora R, Herr W, Pabo CO (1994) Crystal structure of the Oct-1 POU domain bound to an octamer site: DNA recognition with tethered DNA-binding modules. Cell 77: 21–32

Kruse M, Mikoc A, Cetkovic H, Gamulin V, Rinkevich B, Müller IM, Müller WEG (1994) Molecular evidence for the presence of a developmental gene in the lowest animals: identification of a homeobox-like gene in the marine sponge *Geodia cydonium*. Mech Age Dev 77: 43–54

Kurosawa Y, Hashimoto K (1996) The immunoglobulin superfamily: where do invertebrates fit in? Adv Comp Envir on Physiol 23: 151–184

Lake JA (1990) Origin of the Metazoa. Proc Natl Acad Sci USA 87: 763–766

Leiting B, De Francesco R, Tomei L, Cortese R, Otting G, Wüthrich K (1993) The three-dimensional NMR-solution structure of the polypeptide fragment 195–286 of the LFB1/HNF1 transcription factor from rat liver comprises a nonclassical homeodomain. EMBO J 12: 1797–1803

Lewis EB (1978) A gene complex controlling segmentation in *Drosophila*. Nature 276: 565–570

Malicki J, Cianetti LC, Peschle C, McGinnis W (1992) A human HOX4B regulatory element provides head-specific expression in *Drosophila* embryo. Nature 358: 345–347

Manak JR, Scott MP (1994) A class act: conservation of homeodomain protein functions. Development (Suppl): 61–71

McGinnis W, Krumlauf R (1992) Homeobox genes and axial patterning. Cell 68: 283–302

McGinnis W, Levine MS, Hafen E, Kuroiwa A, Gehring WJ (1984a) A conserved DNA sequence in homeotic genes of the *Drosophila* Antennapedia and bithorax complexes. Nature 308: 428–433

McGinnis W, Garber RL, Wirz J, Kuroiwa A, Gehring WJ (1984b) A homologous protein coding sequences in *Drosophila* homeotic genes and its conservation in other metazoans. Cell 38: 403–409

McGinnis N, Kuziora MA, McGinnis W (1990) Human Hox-4.2 and *Drosophila* deformed encode similar regulatory specificities in *Drosophila* embryos and larvae. Cell 63: 969–976

Murtha MT, Leckman JF, Ruddle FH (1991) Detection of homeobox genes in development and evolution. Proc Natl Acad Sci USA 88: 10771–10715

Naito M, Ishiguro H, Fujisawa T, Kurosawa Y (1993) Presence of eight distinct homeobox-containing genes in cnidarians. FEBS Lett 333: 271–274

Nüsslein-Volhard C, Wieschaus E (1980) Mutations affecting segment number and polarity in *Drosophila*. Nature 287: 795–801

Ohno S (1970) Evolution by gene duplication. Springer, Berlin Heidelberg New York

Oliver G, Vispo M, Mailhos A, Martinez C, Sosa-Pineda B, Fielitz W, Ehrlich R (1992) Homeoboxes in flatworms. Gene 121: 337–342

Orii H, Agata K, Watanabe K (1993) POU-domain genes in planarian *Dugesia japonica*: the structure and expression. Biochem Biophys Res Commun 192: 1395–1402

Otting G, Qian YQ, Billeter M, Müller M, Affolter M, Gehring WJ, Wüthrich K (1990) Protein-DNA contacts in the structure of a homeodomain-DNA complex determined by nuclear magnetic resonance spectroscopy in solution. EMBO J 9: 3085–3092

Pabo CO, Sauer RT (1992) Transcription factors: structural families and principles of DNA recognition. Annu Rev Biochem 61: 1053–1095

Phelan ML, Rambaldi I, Featherstone MS (1995) Cooperative interactions between HOX and PBX proteins mediated by a conserved peptide motif. Mol Cell Biol 15: 3989–3997

Qian YQ, Billeter M, Otting G, Müller M, Gehring WJ, Wüthrich K (1989) The structure of the Antennapedia homeodomain determined by NMR spectroscopy in solution: comparison with prokaryotic repressors. Cell 59: 573–580

Qian YQ, Furukubo-Tokunaga K, Resendez-Perez D, Müller M, Gehring WJ, Wüthrich K (1994) Nuclear magnetic resonance solution structure of the fushi tarazu homeodomain from *Drosophila* and comparison with the antennapedia homeodomain. J Mol Biol 238: 335–345

Quiring R, Walldorf U, Kloter U, Gehring WJ (1994) Homology of the eyeless gene of *Drosophila* to the small eye gene in mice and aniridia in humans. Science 265: 785–789

Rosenfeld MG (1991) POU-domain transcription factors; pou-er-ful developmental regulators. Genes Dev 5: 897–907

Ruberti I, Sessa G, Lucchetti S, Morelli G (1991) A novel class of plant proteins containing a homeodomain with a closely linked leucine zipper motif. EMBO J 10: 1787–1791

Ruddle FH, Benthey KL, Murtha MT, Risch N (1994) Gene loss and gain in the evolution of the vertebrates. Development (Suppl): 155–161

Schena M, Davis RW (1992) HD-zip proteins: members of an *Arabidosis* homeodomain protein superfamily. Proc Natl Acad Sci USA 89: 3894–3898

Schneitz K, Spielmann P, Noll M (1993) Molecular genetics of aristaless, a prd-type homeobox gene involved in the morphogenesis of proximal and distal pattern elements in a subset of appendages in *Drosophila*. Genes Dev 7: 114–129

Schummer M, Scheurlen I, Schaller C, Galliot B (1992) HOM/HOX homeobox genes are present in hydra (*Chlorohydra viridissima*) and are differentially expressed during regeneration. EMBO J 11: 1815–1823

Scott MP, Weiner AJ (1984) Structural relationships among genes that control development: sequence homology between Antennapedia, Ultrabithorax, and fushi tarazu loci of *Drosophila*. Proc Natl Acad Sci USA 81: 4115–4119

Scott MP, Weiner AJ, Hazelrigg TI, Polisky BA, Pirrotta V, Scalenghe F, Kaufman TC (1983) The molecualr organization of the Antennapedia locus of *Drosophila*. Cell 35: 763–776

Scott MP, Tamkun JW, Hartzell III GW (1989) The structure and function of the homeodomain. Biochem Biophys Acta 989: 25–48

Seimiya M, Ishiguro H, Miura K, Watanabe Y, Kurosawa Y (1994) Homeobox-containing genes in the most primitive metazoa, the sponges. Eur J Biochem 221: 219–225

Seimiya M, Watanabe Y, Kurosawa Y (1997) Identification of POU-class homeobox genes in a freshwater sponge and the specific expression of these genes during differentiation. Eur J Biochem 243: 27–31

Shenk MA, Bode HR, Steele RE (1993) Expression of Cnox-2, a HOM/HOX homeobox gene in hydra, is correlated with axial pattern formation. Development 117: 657–667

Simeone A, D'Apice MR, Nigro V, Casanova J, Graziani F, Acampora D, Avantaggiato V (1994) Orthopedia, a novel homeobox-containing gene expressed in the developing CNS of both mouse and *Drosophila*. Neuron 13: 83–101

Treisman J, Gönczy P, Vashishtha M, Harris E, Desplan C (1989) A single amino acid can determine the DNA-binding specificity of homeodomain proteins. Cell 59: 553–562

Wainright PO, Hinkle G, Sogin ML, Stickel SK (1993) Monophyletic origins of the metazoa: an evolutionary link with fungi. Science 260: 340–342

Wolberger C, Vershon AK, Liu B, Johnson AD, Pabo CO (1991) Crystal structure of a MATα2 homeodomain-operator complex suggests a general model for homeodomain-DNA interactions. Cell 67: 517–528

Yamada S, Hata J, Yamashita S (1993) Molecular cloning of fish *Pit-1* cDNA and its functional binding to promoter of gene expressed in the pituitary. J Biol Chem 268: 24361–24366

Homeobox-Containing Genes in Freshwater Sponges: Characterization, Expression, and Phylogeny

E. Richelle-Maurer[1], G. Van de Vyver[1], S. Vissers[1], and C.C. Coutinho[2]

1 Introduction

The mono- or polyphyletic origin of metazoans has been discussed for a long time because morphological and embryological data are insufficient to discriminate between these alternative hypotheses. The advent of molecular biology and molecular phylogeny with the development of sequencing and the polymerase chain reaction (PCR) has brought some light to this controversy, although molecular data have sometimes led to conflicting interpretations (Field et al. 1988; Lake 1990; Christen et al. 1991; Kobayashi et al. 1993; Wainright et al. 1993; Rodrigo et al. 1994). At present, the monophyly of metazoans is widely accepted, but the phylogenetic position of sponges, considered to be the most primitive metazoan phylum, remains controversial (Müller et al. 1994a; Müller 1995).

Recently, it has been demonstrated that sponges and metazoans share several homologous molecules, such as collagen (Exposito et al. 1990), protein tyrosine kinase (Ottilie et al. 1992), receptor tyrosine kinase (Schäcke et al. 1994), lectins (Gamulin et al. 1994), ubiquitin (Pfeifer et al. 1993; Müller et al. 1994b), *ets* transcription factors (Degnan et al. 1993), and retinoic acid response elements (Biesalski et al. 1992; Imsiecke et al. 1994), which are very different from those, if present, of taxa other than animals (reviewed in Shenk and Steele 1993). These observations and the developmental role of the extracellular matrix in animals (Morris 1993) suggest a monophyletic origin of metazoans, including sponges.

In this context, the investigation of homeobox-containing genes in sponges may provide additional arguments to sustain this hypothesis. Indeed, the highly conserved homeodomain sequence of homeobox proteins is particularly well suited for phyletic analyses (Ruddle et al. 1994).

Homeobox genes were first identified in *Drosophila melanogaster* (Laughon and Scott 1984; Mc Ginnis et al. 1984; Gehring 1987). They form a family of transcription factors that code for proteins with a 60-amino acid conserved

[1] Laboratoire de Physiologie Cellulaire et Génétique des Levures, CP 244, Université Libre de Bruxelles, Bd du Triomphe, 1050 Brussels, Belgium
[2] Instituto de Quimica, Universidade Federal do Rio de Janeiro, Ilha do Fundao, CEP 21945, Rio de Janeiro, Brazil

Progress in Molecular and Subcellular Biology, Vol. 19
W.E.G. Müller (Ed.)

homeodomain of the helix-turn-helix type (Scott et al. 1989; Affolter et al. 1990; Mc Ginnis and Krumlauf 1992). Some members of this gene family are clustered on the chromosomes forming the *Antennapedia (Antp)*-class gene complex, known as HOM-C genes (Laughon et al. 1986). Such clusters have also been found in nematodes (Bürglin et al. 1991; Kenyon and Wang 1991; Wang et al. 1993) and higher metazoans (Akam 1989; Graham et al. 1989; Singh et al. 1991) forming evolutionary conserved HOM-C/HOX-C complexes. There is every indication that high selective constraint occurred among animals with antero-posterior structure to conserve the HOM-C/HOX-C complex (Ruddle et al. 1994). These gene complexes function in a combinatorial way to determine the identity of the segments of the developing embryo. Other members of the homeobox gene family are spread throughout the chromosome. They are thought to be transcription factors involved in cell-type specification.

In fungi and plants, only very divergent homeoboxes have been found. Yeast homeobox-containing genes play a key role in the determination of the mating type in both budding and fission yeast (Shepherd et al. 1984). In the fungal pathogen *Ustilago maydis*, they orchestrate the growth form, either unicellular or filamentous (Schultz et al. 1990). Plant homeoboxes have been identified only recently in *Arabidopsis thaliana* (Ruberti et al. 1991) and in maize (Vollbrecht et al. 1991), where mutations of the *Knotted* gene alter leaf development.

In metazoans, homeobox-containing genes have been found in all phyla investigated so far, ranging from cnidarians to chordates (e.g., Dolecki et al. 1986; Holland and Hogan 1986; Bürglin et al. 1989; Wang et al. 1990; Garcia-Fernandez et al. 1991; Murtha et al. 1991; Schierwater et al. 1991; Oliver et al. 1992; Schummer et al. 1992; Miller and Miles 1993; Pendleton et al. 1993) and, more recently, in sponges. The presence of these genes in porifera is of evolutionary significance, as sponges are considered to be the lowest multicellular animals. They consist of only a few cell types, lack true tissues or organs, and do not display any type of symmetry (reviewed in Simpson 1984). The first evidence of the presence of homeobox-containing genes in sponges was presented at the 4th International Porifera Congress (Amsterdam, April 1993), where we described the isolation of a 110-bp homeobox fragment, *EfH-1*, from the freshwater *Ephydatia fluviatilis* (Coutinho et al. 1994). *EfH-1* is more related to the *Antp*-type than to the fungi/plant-type homeoboxes. It has 62 and 54% identity with the chicken *Chox 7* and the nematode *ceh-9* homeodomains, respectively (Countinho et al. 1994). Since then, several homeobox-containing genes from freshwater sponges, as well as from marine sponges, have been described. *Prox1*, *prox2*, and *prox3* were isolated from *E. fluviatilis* (Seimiya et al. 1994), SHOX from *Geodia cydonium* (Kruse et al. 1994), *SpoxTA1* from *Tethya aurantia*, and *SpoxH1* and *SpoxH2* from *Haliclona* sp. (Degnan et al. 1995). *Prox1*, *prox2*, *prox3*, and SHOX homeoboxes have been completely sequenced, whereas only a 69-bp homeobox fragment has been identified for *SpoxTA1*, *SpoxH1*, and *SpoxH2*. *Prox2* homeodomain shows 62% identity to the *Om(1D)* homeodomain of *D. melanogaster*, whereas *prox1* and *prox3* homeodomains exhibit the highest degree of identity to the *Drosophila NK-3* and *msh* homeodomains, 63 and 73%, respectively. *EfH-1* shares 92% similarity with *prox2* but only 51% with *proxl* and 59% with *prox3*. The *SpoxTA1* homeobox fragment has high amino acid-sequence identity with *prox 2* and consequently with *EfH-1*. *SpoxH2* from *Haliclona* sp. is closely related to the *Chox7* homeodomain, whereas *SpoxH1* shows high similarity to *Antp*-like homeobox genes (Degnan et al. 1995). If confirmed, it would be the first time that a HOM/HOX homeobox gene had been identified in sponges (Degnan et al. 1995). However, the presence of HOM/HOX clusters in sponges has not yet been demonstrated.

In the first part of this chapter, we report the identification and characterization of a novel homeobox-containing gene, *EmH-3*, from the freshwater sponge *Ephydatia muelleri*. The expression pattern of *EmH-3* gene is consistent with a role in cell differentiation, especially at the time of gemmule hatching.

In the second part of the chapter, we analyze the phylogenetic relationships of all complete sponge homeodomains sequenced up to now. The constructed phylogenetic tree strongly supports a monophyletic origin for the animal kingdom.

Finally, we propose to consider not only the presence of HOM/HOX cluster-type genes but also of all *Antp*-type related homeobox genes as the characteristic defining (synapomorphy) the kingdom Animalia (Slack et al. 1993).

2 Materials and Methods

2.1 Purification and Labelling of DNA Fragments

The *EfH-1* and the 12 kb *E. muelleri* DNAs were purified from agarose gels using the Gene Clean procedure (BIO 101, USA). They were radioactively labeled with [α-^{32}P] dATP (3.000 Ci/mmol, Amersham) using the Random Primed DNA Labeling kit from Boehringer (Mannheim, Germany).

2.2 Screening of a Sponge Genomic Library

A genomic *E. muelleri* library in λGEM-11 (Exposito and Garrone 1990) was screened with the 110-bp *E. fluviatilis EfH-1* homeobox (Coutinho et al. 1994) as a probe (Sambrook et al. 1989). Phage plaques (5.6×10^4) were transferred to nitrocellulose filters. They were hybridized for 15 h at 65 °C in 3 × SSMC, 5 × Denhardt's solution, EDTA 1.25 mM (pH 8), and 100 µg/ml of denatured salmon sperm DNA with the radioactive probe (2×10^5 to 2×10^6 cpm/ml). The filters were washed in 1 × SSC to 0.1 × SSC at 22 or 50 °C, depending on the rate of radioactive decay after each wash. The positive clones were cut out of the original solid culture and stored at 4 °C in a SM solution with 1% chloroform.

2.3
Random Sequencing of the 12-kb Cloned Sponge DNA

The λGEM-11 clone containing the 12-kb genomic sponge DNA was fragmented by sonication. The generated fragments were treated with Klenow polymerase and with T_4 DNA polymerase in order to fill their extremities. Those fragments ranging from 0.5 to 2 kb were purified (Gene Clean Kit, BIO 101) after electrophoretic separation in 0.7% agarose gel and were inserted into the blunt site of pSeqI, a pGEM7zf(+)-derived vector. After transformation in *E. coli*, these plasmids were screened by colony hybridization (Maas 1983) using *EfH-1* as a probe. A total of 264 purified clones constituted the sequencing library. Eighty-six fragments were sequenced by the dideoxynucleotide chain-termination method (Sanger et al. 1977).

2.4
Sequence Analysis

Computer analyses of nucleotide and amino acid sequences were carried out with the Beckman Microgenie software package (Queen and Korn 1984) and with the National Biomedical Research Foundation (NBRF, Georgetown University) protein data bank.

The ClustalV program (Higgins et al. 1992) was used for the multiple-sequence alignment. The evolutionary distance between two aa sequences was measured by the proportion of different aa between the sequences (p-distance). For the phylogenetic inference, the neighbor-joining method (Saitou and Nei 1987) was chosen, since it has been shown to be independent of whether the evolutionary rate varies with lineage or not (Rzhetsky and Nei 1993), and it is quite efficient in recovering the correct tree (Huelsenbeck 1995). Two reliability tests, the bootstrap method (Felsenstein 1985) and the interior-branch test (Sitnikova et al. 1995), were used to measure the degree of support of the generated trees. The bootstrap test has been shown to be quite conservative, whereas the interior-branch test is known to be less conservative and more appropriate for short sequences (Sitnikova et al. 1995). The distance matrix, the bootstrap test, and the respective neighbor-joining tree were all conducted by using the computer package MEGA (Kumar et al. 1993), whereas the interior-branch test and the corresponding neighbor-joining tree were computed by using the METREE program (Rzhetsky and Nei 1992).

2.5
Sponge Culture

Sponges were grown from asexual buds (gemmules). The gemmules of the species *E. muelleri* and *E. fluviatilis*, strain α and δ (Van de Vyver 1970), were collected from sponges cultivated in a pond near Brussels (Van de Vyver and Willenz 1975). They were stored at 4 °C in M mineral medium (Rasmont 1961) until use. Before being incubated, gemmules were treated with 1% perhydrol to reduce microbial contaminants included in the coat (Rozenfeld and Curtis 1980). They were cultivated in sterile M medium at 20 °C, in petri dishes for different incubation times.

At 72 h, about half of the gemmules had hatched. The not fully hatched and the hatched gemmules were treated separately as described below. At 84 h, all gemmules had hatched, and canals and choanocytes became differentiated in some sponges. At 96 h, oscules began to appear, and 12 h later canals and oscules were present in all sponges which were considered as fully functional. From then on, sponges continued to grow and developed their aquiferous system.

2.6 RNA Extraction

Total RNA was extracted by TRIzol Reagent, a monophasic solution of phenol and guanidine isothiocyanate, according to the manufacturer's instructions (Life Technologies) every 12 h, from gemmules to fully functional sponges.

Sponge cultures were treated differently before or after hatching. Before hatching, the gemmule internal cells had not migrated through the micropyle, and therefore the gemmules did not adhere to the substratum. The gemmules were collected and ground in a Potter homogenizer, on ice, in the presence of TRIzol reagent. At the moment of hatching, cells came out of the gemmule, adhered and spread on the substratum. Sponges were then simply scraped and mechanically dissociated by pipetting. The empty gemmules were discarded, and the dissociated cells were recovered by centrifugation (500 g, 10 min, 4 °C) and resuspended in TRIzol reagent.

2.7 RT-PCR Experiments

The primers were selected to match nucleotide sequences at the beginning of the first exon and at the beginning of the homeobox (third exon), in order to differentiate the amplification of contaminating genomic DNA, if present, from that of cDNA (see Fig. 2). The sequence of 5′ primer is 5′-ATGGACAACTGCAGGGGTGA-3′, and that of 3′ primer is 5′-CATTCTCCTATTTTGGAACC-3′. These were supplied by Eurogenetec (Belgium). They were identical to those used by Seimiya et al. (1994) for the amplification of *prox2* from *E. fluviatilis*.

Reverse transcription was performed using the Promega Access RT-PCR System as described in the manufacturer's protocol. This two-enzyme, single-tube system incorporates avian myeloblastosis virus reverse transcriptase (AMV-RT) for first-strand cDNA synthesis and *Thermus flavus* (*Tfl*) DNA polymerase for second-strand cDNA synthesis and DNA amplification. RT-

PCR was conducted in a Pharmacia LKB thermal cycler using Multiply-safecups (Sarstedt) and 10 ng of total RNA diluted into a total of 100 µl of RT-PCR reaction mixture.

The reverse transcription and cycling profiles were as follows: (1) for first-strand cDNA synthesis: 1 cycle, 48 °C, 45 min (reverse transcription); 1 cycle, 94 °C, 2 min (AMV RT inactivation and RNA/cDNA primer denaturation); and (2) 40 cycles for second-strand cDNA synthesis and PCR amplification: 94 °C, 30 s; 55 °C, 1 min; 68 °C, 2 min; and (3) final extension: 1 cycle, 68 °C, 7 min.

Quantification of the respective amounts of RT-PCR amplification products was achieved by electrophoresing serial dilution of these products on a 1% agarose gel.

3 Results and Discussion

3.1 Cloning and Sequencing of the *EmH-3* Homeobox-Containing Gene

The screening of 5.4×10^4 λGEM-11 clones of the *E. muelleri* genomic library with the *E. fluviatilis EfH-1* homeobox as a probe (Coutinho et al. 1994) allowed the identification of a positive clone with an insertion of 12 kb. This fragment was randomly sequenced and a single 180 bp homeobox was identified in the insert close to its 3′ end. The 3′ terminal DNA region containing this homeobox, designated *EmH-3*, was sequenced in both directions over 1.41 kb.

3.2 *EmH-3* Gene Organization

The analysis of the 1.41-kb sequence revealed the presence of a single open reading frame (ORF), 807 nt long, interrupted by two intervening sequences, a putative TATA box and a Cap site 84 and 60 nucleotides upstream of the ATG initiation codon, respectively (Fig. 1). Computer analyses using the Signal PC GENE program indicated possible splice sites on the base of the conserved consensus for RNA splicing. The first intron extended either from nt 139 or nt 193 to nt 243, and the second intron from nt 282 to nt 537. The homeobox spanned nt 592 to nt 771 in the third exon, 36 bp from the C-terminal end of the 12-kb fragment. The open reading frame extended to the 3′ end of the fragment without a stop codon. This indicated that part of the C-terminal coding region was missing.

In order to choose the correct limits of the interruptions in the ORF, RT-PCR experiments were carried out with *E. muelleri* and with the α and δ strains of *E. fluviatilis* for comparison, using the same primers for both species (Fig. 2). The *EmH-3* transcript was about 440 bp long, 50 bp longer than that of *E. fluviatilis*, which had 390 bp, the expected *prox2* transcript size according to

```
EmH-3    -98 - ACATAGACTGATAGTAAATATTTAGGCCGATTATTTTGACCAGAACATGA
               I II II II    IIII   IIII    IIIIIII        III
prox2    -72 - AAATTGATTGC---TAAA----TAGGT--TTTATTTT---------ATG-

                                                          MET
EmH-3        - CTGCTAAATTTTCCTTTCACTTCCAATCAGCCCTAATTAGACAGCAACAT - +2
               IIIII      II  I  I II IIIIIIIIIII I II IIIIIIIII
prox2        - CTGCT------TCAATA-ATTTTCAATCAGCCCTAGTCAGGCAGCAACAT - +2
                                                          MET

                AspAsnCysArgGlyAspLysLysProLeuLeuSerThrAsnGlnGln
EmH-3        - GGACAACTGCAGGGGTGATAAAAAACCTCTACTTTCAACAAACCAACAAT - +52
               IIIIIIIIIIIIIIIIII IIIIIIIIIIIIII IIIIIIIIIIIIIIII
prox2        - GGACAACTGCAGGGGTGAAAAAAAAACCTCTACTGTCAACAAACCAACAAT - +52
                AspAsnCysArgGlyGluLysLysProLeuLeuSerThrAsnGlnGln

               SerPheArgIleAspAsnLeuLeuThrArgLysValIleGluGlnGlnGln
EmH-3        - CTTTCAGAATAGACAACCTTCTTACACGAAAGGTCATAGAGCAGCAACAG - +102
               IIIIIIIIIIII IIIIIIIIIIIIIIIIIIIIIIIIIIIIIIIIIIIII
prox2        - CTTTCAGAATAGATAACCTTCTTACACGAAAGGTCATAGAGCAGCAACAG - +102
               SerPheArgIleAspAsnLeuLeuThrArgLysValIleGluGlnGlnGln

               GlnProAspHisTyrThrMetTyrProProSerLysValGluAsnHisAsp
EmH-3        - CAACCCGATCACTACACCATGTATCCACCATCCAAAGTGGAAAACCATGA - +152
               IIIIIIIIIIIIIIIIIII III IIIIIIIIII II IIIIIII I I
prox2        - CAACCCGATCACTACACCATATATTCACCATCCAAGGTAGAAAACCGTAA - +152
               GlnProAspHisTyrThrIleTyrSerProSerLys

               IleLeuSerLeuThrThrGlyProSerHisAspAspMet
EmH-3        - TATTTAAGCTTGACCACAGGCCCATCTCATGATGACATGGTGAGTGACC - +202
               IIIIII I IIIIIIIIII IIIIIIIIIII         IIII I  II
prox2        - TATTTTTAACTTGACCACAAGCCCATCTCATG--------GTGAAT--CC - +192

                                                           IleSer
EmH-3        - CACAATAGGATTTTTCTTCTTAGAACTGTTACTCTCTTTT-AGATTTCTG - +251
                   I II      III II  IIIIIII IIIIIII I  IIIIIIIII
prox2        - ----AAAG------TCT-CT--GAACTGTCACTCTCTCTGCAGATTTCTG - +229
                                                           IleSer

               AspGlyThrGluIleTyrGluGlnGlyArg
EmH-3        - ATGGAACAGAGATATATGAACAAGGCAGAGGTACATTGTGA-TGGACCAC   +300
               IIIIIII IIIIIIIIIIIII IIIIIIIIIIIIIIIII  III  II
prox2        - ATGGAACCAAGATATATGAACATGGCAGAGGTACATTGTGGCTGG--CAT - +277
               AspGlyThrLysIleTyrGluHisGlyArg

EmH-3        - ATG-CAAATGA---GGGAATAATGTG--TCAC-TGAA-TAAAAATCACAA - +342
               III IIIIIII    I I   II  I   IIII II   I  I  I    I
prox2        - ATGTCAAATGATCAGTGCTGAACATAAATCACATGCCCTCTAGCTATGTA - +327

EmH-3        - CCCACTAA-AATTGAATCACATGTTTTTGTAG---TCATTCTTACAAGTG - +388
               II   II  II    II  I  II III  II      I  II  IIIIIII
prox2        - CCTTTTATGAAGATAAATATTTGATTTGTTAAAACTATTTAATACAAGTA - +377

EmH-3        - ATATTAAATGTTCCCGCATGACATATGTACATTTTTTATGCAATATATAC - +438
                 IIIIIIIIII    I III II I  I III    I I  I II  II I
prox2        - TCATTAAATGTTTTTGTATG-CAGACATTCATAAAT-AGACTATC-ATGC - +424
```

Fig. 1. Nucleotide sequence and deduced amino acid sequence (three-letter code) of the *EmH-3* homeobox-containing gene from *E. muelleri.* (GenBank accession no U97664) Comparison with *prox2* from *E. fluviatilis* (Seimiya et al. 1994). Nucleotides are *numbered* in the *right margin.* The translation start site, the putative TATA box, the conserved nucleotides (GT and AG) at the ends of each intron, and the primers used for RT-PCR are *underlined.* The homeobox is *framed.* In the comparison with *prox2, gaps* were introduced to optimize the alignment (continued on page 164)

```
EmH-3   - ACATATGTAAGTCACATTGATTTTACTCTGAGTTG--TATAGAATT-TTA  - +485
          ||| ||| |  | | ||  |     | |    |||  |||||  || |||
prox2   - ACA-ATGCA--TTATATACAACACAATGAATATTGCTTATAGGTTTGTTA  - +471

EmH-3   - TTCAAAGCCT--TAACCACATTTGCCAGCACCAAGTTACTTGCTTTTCAT  - +533
          | ||  || |  ||| | |||||||| |       ||||||||| |||  |
prox2   - TACATTGCTTCATAAGCTCATTTGCCGG------GTTACTTGCATTTGTT  - +515

                 GluSerThrSerSerThrSerGlyAsnAspAlaGluAspAspLeu
EmH-3   - TTCAGAGTCCACATCTAGTACAAGTGGAAATGATGCTGAGGACGACCTTT  - +583
          |||||||||||||||||||||||||||||||||| ||||| || |||||||
prox2   - TTCAGAGTCCACATCTAGTACAAGTGGAAATGACGCTGACGATGACCTTT  - +565
                 GluSerThrSerSerThrSerGlyAsnAspAlaAspAspAspLeu

          LeuThrArgArgLysLysAlaArgThrAlaPheSerArgGluGlnValAla
EmH-3   - TGACTAGAAGAAAGAAAGCGAGGACAGCCTTTTCTCGAGAACAAGTTGCT  - +633
          ||||||||||||||||||| || || ||||||||||| |||||||||||
prox2   - TGACTAGAAGAAAGAAAGCAAGAACTGCCTTTTCTCGCGAACAAGTTGCA  - +615
          LeuThrArgArgLysLysAlaArgThrAlaPheSerArgGluGlnValAla

          GluLeuGluLysLysPheGlnAspLysLysTyrLeuSerSerAlaGluArg
EmH-3   - GAATTGGAAAAAAAAATTTCAAGACAAAAAGTATCTTTCTTCAGCAGAGAG  - +683
          |||||||||||||||||||| ||||||||||||| |||||   | |||||
prox2   - GAATTGGAAAAAAAAATTTCAAAACAAAAAGTATCTATCTTCTACTGAGAG  - +665
          GluLeuGluLysLysPheGlnAsnLysLysTyrLeuSerSerThrGluArg

          GlyGluLeuAlaGluLysLeuLysLeuSerAspMetGlnValLysThr
EmH-3   - AGGTGAACTAGCAGAAAAACTTAAACTTTCTGACATGCAAGTGAAAACAT  - +733
           ||||||||||||| || ||||||||||||||||||||||||||||||||
prox2   - GGGTGAACTAGCAGAGAAGCTTAAACTTTCTGACATGCAAGTGAAAACAT  - +715
          GlyGluLeuAlaGluLysLeuLysLeuSerAspMetGlnValLysThr

          TrpPheGlnAsnArgArgMetLysTyrLysArgGlnSerGluGluThrGlu
EmH-3   - GGTTCCAAAATAGGAGAATGAAATACAAAAGGCAGTCTGAAGAGACAGAG  - +783
          |||||||||||||||||||||| ||||| ||||||||||||||||| |||
prox2   - GGTTCCAAAATAGGAGAATGAAGTACAAGAGGCAGTCTGAAGAGACTGAG  - +765
          TrpPheGlnAsnArgArgMetLysTyrLysArgGlnSerGluGluThrGlu

          MetGluMetLysSerProLysTyr
EmH-3   - ATGGAAATGAAATCACCCAAATAC                            - +807
          ||||| ||||||||| |||||||
prox2   - ATGGAGATGAAATCATCCAAATATGGGTCTTTTGTATATCGTGGAGTGGC  - +815
          MetGluMetLysSerSerLysTyrGlySerPheValTyrArgGlyValAla

prox2   - ACATTCTATGTTCAATGCAGTG                              - +837
          HisSerMetPheAsnAlaVal
```

Fig. 1. *(Continued)*

Seimiya et al. (1994). This clearly indicated that the first exon/intron junction was located at position nt 193 and not at nt 139 as in *E. fluviatilis* (Fig. 1). The restriction profile (not shown) of RT-PCR products digested with *Hind* III were in agreement with known *Hind* III restriction sites present in *EmH-3* and *prox2* genes.

The schematic exon/intron structure of the *EmH-3* gene is shown in Fig. 3. It is very similar to that of the *prox2* gene of *E. fluviatilis*, except that the *EmH-3* first exon is 54 bp longer than that of *prox2*. The *EmH-3* third exon is incomplete, but if we assume that this exon has the same length in both species, the missing part of *EmH-3* gene only represents 9.2% corresponding to 16 aa.

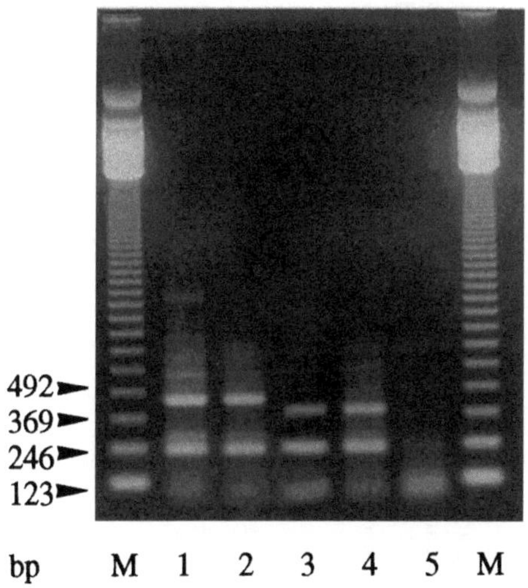

Fig. 2. Characterization of the *EmH-3* transcripts. Amplified products of RT-PCR (5 µl of the reaction electrophoresed in a 1% agarose gel) of total RNA isolated from 5-day-old sponges. *Lanes* 1 and *2*: *E. muelleri*; *lane 3*: *E. fluviatilis* strain α; *lane 4*: *E. fluviatilis* strain δ; *lane 5*: negative control without RNA template; *M* molecular-size marker

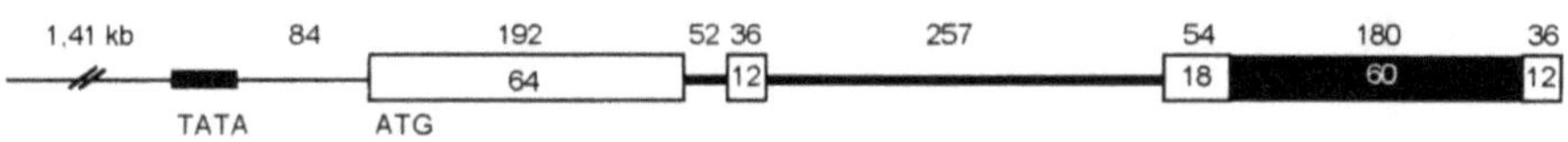

Fig. 3. Structural organization of the *EmH-3* homeobox-containing gene. Exons are represented by *open boxes*, the homeobox by a *solid box*, and introns by *solid lines*. The last exon is incomplete, as the C-terminal end was missing in the 1.41-kb fragment and could not be sequenced. *Numbers above the boxes* indicate the size in bp of exons and introns, *numbers inside the boxes*, the size in aa of the exons

If so, the *EmH-3* homeobox would then lie in the 3′ terminal exon, as is the case in most known homeoboxes (Scott et al. 1989).

The comparative analysis of *EmH-3* and *prox2* nt sequences (Seimiya et al. 1994) showed that they shared 75% identity. The identity was particularly high in the translated regions as compared with untranslated regions. However, a conserved DNA island was also observed between nt 139 and nt 184, at the 3′ extremity of the first predicted exon of *E. muelleri* and the 5′ extremity of the first predicted intron of *E. fluviatilis*.

3.3 Analysis of the *EmH-3* Homeobox Protein

The deduced aa sequence of *EmH-3* homeobox protein displayed the homeodomain signature LAEKLKLSDMQVKTWFQNRRMKYK. It possessed characteristic features of homeodomains (Scott et al. 1989; Laughon 1991; Kappen and Ruddle 1993) as shown in Fig. 4: 11 of the 12 highly conserved aa residues among arthropods and vertebrates (Ile 45 replaced by Val), and 16 of the 21 aa involved in base and phosphate contacts specific to the *Antp* homeodomain (Arg 3 replaced by Lys, Ile 34 by Leu, Ala 37 by Lys, Ile 45 by Val, Ile 47 by Thr).

A

```
        10        20        30        40        50        60
  B          H         PH   P  H  H H         HHBB PBP P
RKKARTAFSREQVAELEKKFQDKKYLSSAERGELAEKLKLSDMQVKTWFQNRRMKYKRQS
   ▲      ▲  ▲   ▲                  ▲          ▲▲ ▲ ▲ ▲  ▲
```

B

```
EmH-3   MDNCRGDKKPLLSTNQQSFRIDNLLTRKVIEQQQQPDHYTMYPPSKISDGTEIYEQGRESTSSTSGNDAEDDLLTR
Prox2   ------E*--------------------------------I*-S--------K---H*-------------D------

EmH-3   RKKARTAFSREQVAELEKKFQDKKYLSSAERGELAEKLKLSDMQVKTWFQNRRMKYKRQSEETEMEMKSPKY
Prox2   --------------------N------T*--------------------------------------------
```

Fig. 4. A *EmH-3* homeodomain. The 11 highly conserved aa are marked by *triangles*. The four invariant aa among nonyeast homeodomains are at positions 48, 49, 51, and 53 (Gehring et al. 1990). The positions corresponding to the hydrophobic core (*H*), base (*B*), and phosphate (*P*) aa contact similar to the *Antp* structure (Laughon 1991) are indicated. The four α-helices are *underlined* **B** Comparison of the *EmH-3* and *prox2* homeoboxproteins. *Hyphens* correspond to identical aa, and aa marked with an *asterisk* to conservative substitutions. The homeodomain is *underlined*. The 18 supplementary aa of *EmH-3* first exon are not considered in the comparison

Homology searches using the NBKF data bank showed that the putative *EmH-3* homeodomain shared 61.5% identity with the *Drosophila OM(1D)* homeodomain, but the closest identity, 94%, was found with the *prox2* homeodomain of *E. fluviatilis*. *EmH-3* and *prox2* putative homeobox proteins exhibited a very high degree of identity (85%), reaching as much as 94% when the proposed *EmH-3* supplementary 18 aa residues of the first exon (nt 139-nt 193) were not taken into account. Moreover, the similarities with *prox2* were not restricted to the homeodomain as is the case in most homeoproteins but extended to the variable part of the protein. This degree of homology has also been observed by Southern hybridization (Richelle-Maurer et al. 1996a). These data suggested that *EmH-3* and *prox2* genes could have similar functions and that there have been functional constraints to conserve the aa sequence in both species as proposed by Schierwater et al. (1991) for the cnidarians *Eleutheria dichotoma* and *Hydractinia symbiolongicarpus*. However, the differential splicing due to alternative use of two splice donor sites in the N-terminal region resulted in proteins that may potentially have different functional roles or different target cells (Cho et al. 1988; O'Connor et al. 1988; Wang et al. 1990). More work on the temporal and spatial expression of *EmH-3* and *prox2* and their target cells would be needed to confirm either of these hypotheses.

3.4
Expression of the *EmH-3* Gene in the Course of Development

RT-PCR experiments demonstrated that the *EMH-3* gene was differentially expressed at various stages of development, going from gemmule germination to the completion of a functional sponge. Six of the 12 studied stages are shown on Fig. 5. Almost undetectable in gemmules before incubation, *EmH-3* transcripts were already present 12 h after the beginning of incubation. Their amount increased slightly during the following 12 h and remained nearly constant until hatching (72 h). At that time, a high level of expression, approximately fourfold the 24-h level, was observed in cells issued from completely or partially hatched gemmules. This level of maximal expression was maintained for at least 24 h. Afterward the quantity of transcripts decreased to about one third of the maximal value and appeared to stabilize in the course of development.

Our results differ from those of the only other study of sponge gene expression to date. Seimiya et al. (1994), using the RT-PCR method, showed that *prox1* and *prox2* genes from *E. fluviatilis* were expressed at all stages of differentiation from germination of the gemmule to formation of the choanocyte chambers. However, we have shown, also using the RT-PCR method and the same primers, that *prox2* was differentially expressed in gemmules and fully functional sponges as shown for *E. muelleri* (Richelle-Maurer et al. 1996b).

The enhancement of the expression at the moment of hatching suggests that the *EmH-3* gene is particularly involved at that stage of development, i.e., when

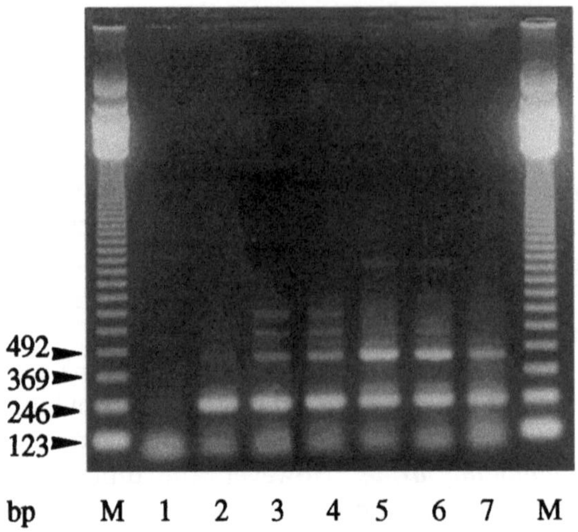

Fig. 5. Expression of the *EmH-3* gene in the course of development. Amplified products of RT-PCR of total RNA isolated at different stages, from gemmules to the formation of fully functional sponges. Developmental stages are expressed as hours after incubation at 20 °C, in mineral M medium. *Lane 1* Negative control without RNA template; *lane 2* 0 h (gemmules before incubation); *lane 3* 12 h; *lane 4* 24 h; *lane 5* 72 h (*hatching*); *lane 6* 96 h; *lane 7* 132 h; *M* molecular-size marker. RT-PCR products were analyzed by electrophresing 5 µl of the reaction in a 1% agarose gel

cell differentiation is at its height. Indeed, in the gemmule before hatching, the only observable morphological change is the division of the binucleate thesocytes into mononucleate archaeocytes. This process lasts from about 12 to 48 h for incubation at 20 °C (reviewed in Simpson 1984). In contrast, at hatching there is a burst of differentiation and metabolic processes. All sponge cells differentiate from the archaeocytes. Peaks of expression are observed, for instance, for actin or collagen genes (Ducy 1993; Lu 1994). The first cells to differentiate after the migration of the archaeocytes through the open micropyle are the pinacocytes, which constitute the external epithelium of the sponge, and the sclerocytes, which produce its spicules (reviewed in Simpson 1984). Finally, the choanocytes are formed by several successive mitoses of the archaeocytes (Rozenfeld and Rasmont 1976), and afterward there is a slowing down of these differentiation processes. Concurrently, we observed a decrease in the *EmH-3* expression. The maintenance of this expression throughout the sponge's life, although at a lower level, is probably correlated to a continuous differentiation process, in particular that of choanocytes, which are continuously replaced.

3.5 Phylogeny of Sponge Homeobox-Containing Genes

To analyze evolutionary relationships of sponge and other organisms' homeobox-containing genes, a phylogenetic tree based on aa sequences data was constructed using the neighbor-joining method. Amino acid sequence data were chosen rather than nucleotide sequences, because synonymous nucleotide substitutions were apparently saturated in most gene comparisons, and this would have introduced noise in constructing phylogenetic trees (Russo et al. 1996). The homeodomain sequences from sponges and from a variety of organisms were aligned using the ClustalV computer package (Higgins et al. 1992).

The comparison of *EmH-3* homeodomain with all known sponge homeodomains revealed the highest identity and similarity with *prox2* from *E. fluviatilis* (97/98%), followed by *SpoxTA1* from *T. aurantia* (87/96%), *prox3* and *prox1* from *E. fluviatilis* (57/70% and 50/70%, respectively), and *SpoxH2* and *SpoxH1* from *Haliclona* sp. (57/70% and 48/52%, respectively). The similarity with SHOX homeodomain was very low (17%), even lower than that found between *EmH-3* and, for instance, *Hydra* homeodomains. The great similarity to *SpoxTA1* isolated from a very primitive sponge, the tetractinomorph *T. aurantia*, is particularly interesting because it suggests that an *EmH-3*-like *gene* could be widespread among porifera and could represent one of their ancestral Hox genes.

On the basis of the percentage of sequence identity with other homeodomains, the sponge homeodomains can be divided into five different classes: the *BarH* class (*EmH-3, EfH-1, prox2,* and *SpoxTA1*), the *NK-3* class (*prox1*), the *msh* class (*prox3*), the *Cdx* class (*SpoxH2*), and possibly the *Antp*-class including *SpoxH1*. However, the phylogenetic relationships among sponge homeodomains and other homeodomains, in particular those of yeast/fungi, have not yet been investigated.

Taking advantage of the homeodomain classes established by Kappen et al. (1993), we aligned *prox1, prox2, prox3, EmH-3,* and SHOX homeodomains with *Antp, msh, NK-3,* members of the *BarH* class, and with the yeast/fungi homeodomains: *yox1, PHO2, MATa1, MATa2, pi,* and *b2* (Schultz et al. 1990; Kaufmann 1993; Kappen et al. 1993). *SpoxH1, SpoxH2* and *SpoxTA1* were not considered, as they are only the internal part of a homeodomain (23 aa).

Most fungi/yeast sequences were so divergent that it was not possible to obtain a reliable alignment using all sequences (maximum p distance of 0.9). The only fungi/yeast sequence that showed a fairly high similarity (p distance of 0.7) with the metazoan sequences was *yox1*. Therefore, it was used as the outgroup, and the others were excluded from the phylogenetic analysis. Similarly, sponge SHOX gene was excluded, since it was even more distant from the classic *Antp*-class than the fungi/yeast's most divergent homeobox, *b2* (results not shown).

The constructed tree is presented in Fig. 6. The interior branch test on the neighbor-joining tree showed a relatively high value (75% probability) grouping the *EmH-3* and *prox2* homeodomains with *BarH1, Om(1D), BarH2, Cnox3, Barx1, Hox11,* and *tcl-3* homeodomains.

Prox1 and *prox3* homeodomains clustered with a high probability value to *Drosophila msh* and *NK-3* homeodomains, respectively. This is in agreement with the percentage of identity found between these homeodomains by Seimiya et al. (1994). The relative position of *msh* and *NK-3* classes was not statistically well supported, and thus their relationship with the *Antp*-class could not be rigorously evaluated.

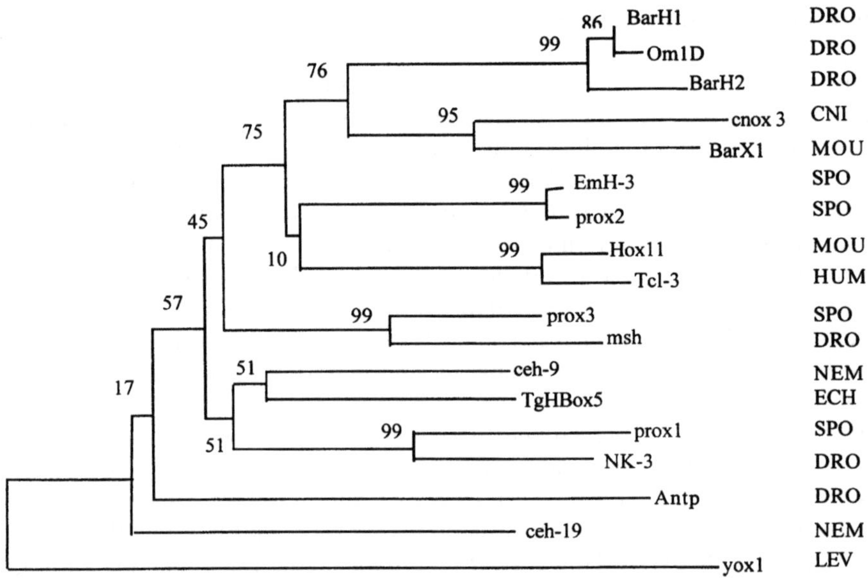

Fig. 6. The phylogenetic tree constructed by neighbor-joining method (Saitou and Nei 1987). Abbreviations: *CNI* cnidarian; *DRO Drosophila*; *ECH* echinoderm; *LEV* yeast; *HUM* human; *MOU* mouse; *NEM* nematode; *SPO* sponge. The data were obtained from Kappen et al. (1993; *Antp*, *BarH1*, *BarH2*, *ceh-9*, *ceh-19*, *cnox 3*, *TgHBox5*, *Hox11*, *msh*, *NK-3*, *Om(1D)*, *tcl-3*), from Seimiya et al. (1994; *prox1*, *prox2*, *prox3*), from Kaufmann (1993; *yox1*), and from Tissier-Seta et al. (1995; *Barx1*). *Numbers* at each branch indicate the result of the interior-branch test

The low probability of the interior branch test on the remaining deep interior branch indicated that the 60-amino acid homeodomain was too short, and there was not enough phylogenetic information to resolve all the interior branches of the tree. There was a clear need for further data to assess the other phylogenetic relationships of this tree.

EmH-3 and all sponge homeodomains analyzed are members of homeobox-class genes that, in nematodes and higher metazoans, are not clustered on the chromosome. In higher metazoans these unclustered genes control cell differentiation, rather than specify regional patterns, as does the HOM/HOX cluster. They have persisted throughout evolution, presumably because they work in a network, and because they are essential for all multicellular animals and could therefore not have been easily eliminated individually. The exception could be the sponge *SpoxH1* gene, a member of the *Antp*-class that in higher metazoans has clustered genes (Degnan et al. 1995). The presence of an *Antp*-class gene in porifera is noteworthy, as the *SpoxH1* gene could represent the most primitive gene of the homeobox HOM/HOX complex. This would lead to a reappraisal of Ruddle's hypothesis that *Hox* cluster gene system may represent a special case involved in axial patterning that is a feature of all phyla beyond sponges (Ruddle et al. 1994).

According to several authors (Murtha et al. 1991; Pendelton et al. 1993; Ruddle et al. 1994), the number of *Hox* genes in an organism would be correlated to its body-plan complexity, which can simply be evaluated by its cell-type number (Valentine 1994). This implies that sponges, being the simplest multicellular organisms, would only have a few homeobox genes, and would probably not be organised in clusters. If the latter existed, they would be very short, comprising only a few genes. Homeobox gene clusters are thought to have arisen early in animal evolution by duplication of an ancestral homeobox gene (Kappen and Ruddle 1993; Kappen et al. 1993; Ruddle et al. 1994). The earlier duplication even would probably have produced more than one *Antp*-type gene in the ancestral cluster (Graham et al. 1989; Zhang et al. 1996), but these hypotheses have not yet been completely verified. Therefore, further experimental studies of the structure, function, and expression patterns of homeobox-containing genes in sponges will be essential to understand the evolutionary history of Hox genes.

4 Conclusions

Our experiments clearly demonstrate that there is a temporal expression of *E. muelleri*, *EmH-3*, homeobox-containing gene during gemmule germination and hatching. They provide strong evidence for the involvement of this gene in cell differentiation, especially at the moment of hatching. *EmH-3* displays characteristic features common to animal homeodomains and, as shown by the phylogenetic analysis, is much more closely related to the homeodomains of metazoans, in particular to the *Antp*-class, than to those of yeast/fungi. These results bring additional arguments for the existence of a common ancestor between sponges and other metazoans not shared with fungi, plants, or protists, unifying the phylum metazoa into a monophyletic group. They lead us to propose including in the definition of the zootype all homeobox-containing genes related to the *Antp*-class, whether they are found in clusters or not.

Acknowledgements. We are grateful to Professor R. Garrone for providing the *E. muelleri* genomic library. We thank Dr C. Russo for carrying out the phylogenetic analyses.

References

Affolter M, Schier A, Gehring WJ (1990) Homeodomain proteins and the regulation of gene expression. Curr Opin Cell Biol 2: 485–495

Akam M (1989) *Hox* and HOM: Homologous gene clusters in insects and vertebrates. Cell 57: 347–349

Biesalski HK, Doepner G, Tzimas G, Gamulin V, Schröder HC, Batel R, Nau H, Müller WEG (1992) Modulation of *myb* gene expression in sponges by retinoic acid. Oncogene 7: 1765–1774

Bürglin TR, Finney M, Coulson A, Ruvkun G (1989) *Caenorhabditis elegans* has scores of homeobox-containing genes. Nature 341: 239–243

Bürglin TR, Ruvkun G, Coulson A, Hawkins NC, McGhee JD, Schaller D, Wittman D, Müller F, Waterston RH (1991) Nematode homeobox cluster. Nature 351: 703

Cho KWY, Getz J, Wright CVE, Fritz A, Hardwicke J, De Robertis EM (1988) Differential utilization of the same reading frame in a *Xenopus* homeobox gene encodes two related proteins sharing the same DNA-binding specificity. EMBO J 7: 2139–2149

Christen R, Ratto A, Baroin A, Perasso R, Grell KG, Adoutte A (1991) An analysis of the origin of metazoans, using comparisons of partial sequences of the 28S RNA, reveals an early emergence of triploblasts. EMBO J 10: 499–503

Coutinho CC, Vissers S, Van de Vyver G (1994) Evidence of homeobox genes in the freshwater sponge *Ephydatia fluviatilis*. In: van Soest RWM, Van Kempen ThMG, Braekman JC (eds) Sponges in time and space. Balkema Press, Rotterdam, pp 385–388

Degnan BM, Degnan SM, Naganuma T, Morse DE (1993) The *ets* multigene family is conserved throughout the Metazoa, Nucl Acids Res 21: 3479–3484

Degnan BM, Degnan SM, Giusti A, Morse DE (1995) A *hox*-hom homeobox gene in sponges. Gene 155: 175–177

Dolecki GJ, Wannakrairoj S, Lum R, Wang G, Riley HD, Carlos R, Wang A, Humphreys T (1986) Stage-specific expression of a homeobox-containing gene in the non-segmented sea urchin embryo. EMBO J 5: 925–930

Ducy P (1993) Etude d'isoformes d'actine chez le spongiaire *Ephydatia mülleri* LIEB. Thèse de Doctorat, Université Claude Bernard-Lyon I, France

Exposito JY, Garrone R (1990) Characterization of a fibrillar collagen gene in sponges reveals the early evolutionary appearance of two collagen gene families. Proc Natl Acad Sci USA 87: 6669–6673

Exposito JY, Ouazana R, Garrone R (1990) Cloning and sequencing of a Porifera partial cDNA coding for a short-chain collagen. Eur J Biochem 190: 401–406

Felsenstein J (1985) Confidence limits on phylogenies: an approach using the bootstrap. Evolution 39: 783–791

Field KG, Olsen GJ, Lane DJ, Giovannoni ST, Gheselin MT, Raff EC, Pace NR, Raff RA (1988) Molecular phylogeny of the animal kingdom. Science 239: 748–753

Gamulin V, Rinkevich B, Schäcke H, Kruse M, Müller IM, Müller WEG (1994) Cell adhesion receptors and nuclear receptors are highly conserved from the lowest Metazoa (marine sponges) to vertebrates. Biol Chem 375: 583–588

Garcia-Fernandez J, Baguna J, Salo E (1991) Planarian homeobox genes: cloning, sequence analysis and expression. Proc Natl Acad Sci USA 88: 7338–7342

Gehring WJ (1987) Homeoboxes in the study of development. Science 230: 1245–1252

Gehring WJ, Müller M, Affolter A, Percival-Smith A, Billeter M, Qian YQ, Otting G, Wüthrich K (1990) The structure of the homeodomain and its functional implications. TIG 10: 323–329

Graham A, Papalopulu N, Krumlauf R (1989) The murine and *Drosophila* homeobox gene complexes have common features of organisation and expression. Cell 57: 367–378

Higgins DG, Bleasby AJ, Fuchs R (1992) Clustal V: improved software for multiple sequence alignment. CABIOS 8: 189–191

Holland PWH, Hogan BLM (1986) Phylogenetic distribution of *Antennapedia*-like homeoboxes. Nature 321: 251–253

Huelsenbeck JP (1995) Performance of phylogenetic methods in simulation. Syst Biol 44: 17–48

Imsiecke G, Borojevic R, Custodio M, Müller WEG (1994) Retinoic acid acts as morphogen in freshwater sponges. Invertebr Reprod Dev 26: 89–98

Kappen C, Ruddle FH (1993) Evolution of a regulatory gene family: *HOM/HOX* genes. Curr Opin Genet 3: 931–938

Kappen C, Schughart K, Ruddle FH (1993) Early evolutionary origin of the major homeodomain sequence classes. Genomics 18: 54–70

Kaufmann F (1993) In vitro binding to the leucine tRNA gene identifies a novel yeast homeobox gene. Chromosoma 102: 174–179

Kenyon C, Wang B (1991) A cluster of *Antennapedia*-class homeobox genes in a nonsegmented animal. Science 253: 516–517

Kobayashi M, Takahashi M, Wada H, Satoh N (1993) Molecular phylogeny inferred from sequences of small ribosomal DNA, supports the monophyly of the Metazoa. Zool Sci 10: 827–833

Kruse M, Mikoc A, Cetkovic H, Gamulin V, Rinkevich B, Müller IM, Müller WEG (1994) Molecular evidence for the presence of a developmental gene in the lowest animals: identification of a homeobox-like gene in the marine sponge *Geodia cydonium*. Mech Age Dev 77: 43–54

Kumar SK, Tamura K, Nei M (1993) MEGA: molecular evolutionary genetics analysis, version 1.01. Pennsylvania State University, University Park

Lake JA (1990) Origin of the metazoa. Proc Natl Acad Sci USA 87: 763–766

Laughon A (1991) DNA binding specificity of homeodomains. Biochemistry 30: 11357–11367

Laughon A, Scott MP (1984) Sequence of a *Drosophila* segmentation gene: protein structure homology with DNA-binding proteins. Nature 310: 25–30

Laughon A, Boulet AM, Bermingham JR, Laymon RA, Scott MP (1986) Structure of transcripts from the homeotic *Antennapedia* gene of *Drosophila melanogaster*: two promoters control major protein-coding region. Mol Cell Biol 6: 4676–4689

Lu Q (1994) L'origine des collagènes non-fibrillaires: caractérisation de gènes de collagènes primitifs chez le spongiaire *Ephydatia mülleri*. Thèse de Doctorat, Université Claude Bernard-Lyon I, France

Maas R (1983) An improved colony hybridisation method with significantly increased sensitivity for detection of single genes. Plasmid 10: 296–298

McGinnis W, Krumlauf R (1992) Homeobox genes and axial patterning. Cell 68: 283–302

McGinnis W, Garber RL, Wirz J, Kuroiwa A, Gehring WJ (1984) A homologous protein-coding sequence in *Drosophila* homeotic gene and its conservation in other metazoans. Cell 37: 403–408

Miller DJ, Miles A (1993) Homeobox genes and the zootype. Nature 365: 215–216

Morris PJ (1993) The developmental role of the extracellular matrix suggests a monophyletic origin of the kingdom Animalia. Evolution 47: 152–165

Müller WEG (1995) Molecular phylogeny of Metazoa (animals): monophyletic origin. Naturwissenschaft 82: 321–329

Müller WEG, Müller IM, Gamulin V (1994a) On the monophyletic evolution of the metazoa. Braz J Med Biol Res 27: 2083–2096

Müller WEG, Schröder HC, Müller IM, Gamulin V (1994b) Phylogenetic relationship of ubiquitin repeats in the polyubiquitin gene from the marine sponge *Geodia cydonium*. J Mol Evol 39: 369–377

Murtha MT, Leckman JF, Ruddle FH (1991) Detection of homeobox genes in development and evolution. Proc Natl Acad Sci USA 88: 10711–10715

O'Connor MB, Binari R, Perkins LA, Bender W (1988) Alternative RNA products from the *Ultrabithorax* domain of the bithorax complex. EMBO J 7: 435–445

Oliver G, Vispo M, Mailhos A, Martinez C, Sosa-Pineda B, Fielitz W, Ehrlich R (1992) Homeoboxes in flatworms. Gene 121: 337–342

Ottilie S, Raulf F, Barnekow A, Hanning G, Schartl M (1992) Multiple *src*-related kinase genes, *srk*1–4 in the fresh water sponge *Spongilla lacustris*. Oncogene 7: 1625–1630

Pendleton JW, Nagai BK, Murtha MT, Ruddle FH (1993) Expansion of the *Hox* gene family and the evolution of chordates. Proc Natl Acad Sci USA 90: 6300–6304

Pfeifer K, Frank W, Schröder HC, Gamulin V, Rinkevich B, Batel R, Müller IM, Müller WEG (1993) Cloning of the polyubiquitin cDNA from the marine sponge *Geodia cydonium* and its preferential expression during reaggregation of cells. J Cell Sci 106: 543–554

Queen C, Korn LJ (1984) A comprehensive sequence analysis program for the IBM personal computer. Nucl Acids Res 12: 581–599

Rasmont R (1961) Une technique de culture des éponges en mileu contrôlé. Ann Soc R Zool Belg 91: 147–155

Richelle-Maurer E, Kucharczak J, Van de Vyver G, Vissers S (1996a) Southern-blot hybridisation, a useful technique in freshwater taxonomy. Bull Inst R Sci Nat Belg 66: 227–229

Richelle-Maurer E, Coutinho CC, Kucharczak J, Vissers S, Van de Vyver G (1996b) Spatial and temporal expression of homeobox-containing genes in freshwater sponges. Int Conf Sponge Science, Molecular aspects of Sponge Science, 12–16 March 1996, OTSU, Japan

Rodrigo AG, Bergquist PR, Bergquist PL, Reeves RA (1994) Are sponges animals? An investigation into the vagaries of phylogenetic inference. In: van Soest RWM, Van Kempen ThMG, Braekman JC (eds) Sponges in time and space. Balkema Press, Rotterdam, pp 47–54

Rozenfeld F, Curtis ASG (1980) A technique for sterile culture of fresh water sponges. Experientia 36: 371–373

Rozenfeld F, Rasmont R (1976) Hydroxyurea: an inhibitor of the differentiation of choanocytes in fresh water sponges and a possible agent for the isolation of embryonic cells. Differentiation 7: 53–60

Ruberti I, Sessa G, Lucchetti S, Morelli G (1991) A novel class of plant proteins containing a homeodomain with a closely linked leucine zipper motif. EMBO J 10: 1787–1791

Ruddle FH, Bartels JL, Bentley KL, Kappen C, Murtha MT, Pendleton JW (1994) Evolution of *HOX* genes. Annu Rev Genet 28: 423–442

Russo CAM, Takezaki N, Nei M (1996) Efficiencies of different genes and different tree-building methods in recovering a known vertebrate phylogeny. Mol Biol Evol 13: 525–536

Rzhetsky A, Nei M (1992) A simple method for estimating and testing minimum-evolution trees. Mol Biol Evol 9: 945–967

Rzhetsky A, Nei M (1993) Theoretical foundation of the minimum-evolution method of phylogenetic inference. Mol Biol Evol 10: 1073–1095

Saitou N, Nei M (1987) The neighbor-joining method: a new method for reconstructing phylogenetic trees. Mol Biol Evol 4: 406–425

Sambrook J, Fritsch EF, Maniatis T (1989) Molecular cloning, a laboratory manual. Cold Spring Harbor Lab Press, Cold Spring Harbor, New York

Sanger F, Nicklen S, Coulson AR (1977) DNA sequencing with chain-terminating inhibitors. Proc Natl Acad Sci USA 74: 5463–5467

Schäcke H, Schröder HC, Gamulin V, Rinkevich B, Müller IM, Müller WEG (1994) Molecular cloning of a tyrosine kinase gene from the marine sponge *Geodia cydonium*: a new member belonging to the receptor tyrosine kinase class II family. Mol Membr Biol 11: 101–107

Schierwater B, Murtha M, Dick M, Ruddle FH, Buss LW (1991) Homeoboxes in cnidarians. J Exp Zool 260: 413–416

Schultz B, Banuett F, Dahl M, Schlesinger R, Schäfer W, Martin T, Herskowitz I, Kahmann R (1990) The *b* alleles of *U. maidis*, whose combinations program pathogenic development, code for polypeptides containing a homeodomain-related motif. Cell 60: 295–306

Schummer M, Scheurlen I, Schaller C, Galliot B (1992) HOM/HOX homeobox genes are present in hydra (*Chlorohydra viridissima*) and are differentially expressed during regeneration. EMBO J 11: 1815–1823

Scott MP, Tamkun JW, Hartzell GW (1989) The structure and function of homeodomain. Biochim Biophys Acta 989: 25–48

Seimiya M, Ishiguro H, Miura K, Watanabe Y, Kurosawa Y (1994) Homeobox-containing genes in the most primitive Metazoa, the sponges. Eur J Biochem 221: 219–225

Shenk MA, Steele RE (1993) A molecular snapshot of the metazoan "Eve". TIBS 18: 459–463

Shenk MA, Bode HR, Steele RE (1993) Expression of *Cnox-2*, a HOM/HOX homeobox gene in hydra, is correlated with axial pattern formation. Development 117: 657–667

Shepherd JCW, McGinnis W, Carrasco AE, De Robertis EM, Gehring WJ (1984) Fly and frog homeo domains show homologies with yeast mating type regulatory proteins. Nature 310: 70–71

Simpson TL (1984) The cell biology of sponges. Springer, Berlin Heidelberg New York

Singh G, Kaur S, Stock JL, Jenkins NA, Gilbert DJ, Copeland NG, Potter SS (1991) Identification of 10 murine homeobox genes. Proc Natl Acad Sci USA 88: 10706–10710

Sitnikova T, Rzhetsky A, Nei M (1995) Interior-branch and bootstrap tests of phylogenetic trees. Mol Biol Evol 12: 319–333

Slack JMW, Holland PWH, Graham CF (1993) The zootype and the phylotypic stage. Nature 361: 490–492

Tissier-Seta JP, Mucchielli ML, Mark M, Mattei MG, Goridis C, Brunet JF (1995) Barxl, a new mouse homeodomain transcription factor expressed in cranio-facial ectomesenchyme and the stomach. Mech Dev 5: 13–15

Valentine JW (1994) Late Precambrian bilaterians: grades and clades. Proc Natl Acad Sci USA 91: 6751–6757

Van de Vyver G (1970) La non confluence intraspécifique chez les spongiaires et la notion d'individu. Ann Embryol Morphog 3: 251–262

Van de Vyver G, Willenz Ph (1975) An experimental study of the life cycle of the fresh water sponge *Ephydatia fluviatilis* in its natural surroundings. Wihelm Roux's Arch 177: 41–52

Vollbrecht E, Veit B, Sinha N, Hake S (1991) The developmental gene *Knotted-1* is a member of a maize homeobox gene family. Nature 350: 241–243

Wainright PO, Hinkle G, Sogin ML, Stickel SK (1993) Monophyletic origins of the Metazoa: an evolutionary link with fungi. Science 260: 340–342

Wang BB, Müller-Immergluck MM, Austin J, Robinson NT, Chisholm A, Kenyon C (1993) A homeotic gene cluster patterns the anteroposterior body axis of *C. elegans*. Cell 74: 29–42

Wang GVL, Dolecki GJ, Carlos R, Humphreys T (1990) Characterization and expression of two sea urchin homeobox gene sequences. Dev Genet 11: 77–87

Zhang J, Nei M (1996) Evolution of Antennapedia-class homeobox genes. Genetics 142: 295–303

Early Evolution of the Metazoa: An Inference from the Elongation Factor-1α

M. Kobayashi and N. Satoh[1]

1 Introduction

The animal kingdom consists of more than thirty phyla. Based upon extensive studies of comparative anatomy, embryology, and paleontology, metazoans have been divided into three major taxonomic units, namely, animals with extremely undifferentiated tissues (poriferans), those with two embryonic layers and radial symmetry (diploblasts), and those with three embryonic layers and bilateral symmetry (triploblasts). Poriferans are sometimes included among diploblasts, because they have two cell layers. In poriferans, the choanocytes and pinacocytes form tissues to some extent, the choanoderm and pinacoderm, respectively. Furthermore, basement membrane-like structures such as the lamina reticularis (Harrison et al. 1990) and the basal lamina (Pedersen 1991) have been found in sponges. Therefore, we regarded the diploblasts as a unit that consists of three phyla, the Porifera, Cnidaria, and Ctenophora, in addition to two phyla of uncertain affinity, the Placozoa and Mesozoa. The phylogenetic status of these animals has been a matter of interest in regard to the origin and early evolution of the metazoans (Hanson 1977; Ax 1989; Brusca and Brusca 1990; Willmer 1990; Schram 1991; Ruppert and Barnes 1994; Nielsen 1995). Three major unresolved issues relating to the phylogenetic status of diploblasts are: (1) whether all the metazoans, including diploblasts and triploblasts, are monophyletic; (2) whether the diploblasts are monophyletic or paraphyletic; and (3) how the three diploblast phyla are related.

From the morphological standpoint, several authors have proposed that metazoans are polyphyletic (Nursall 1962; Hanson 1977; Inglis 1985). Because of their simple body organization and the lack of many metazoan characteristics, such as true tissue differentiation and synaptic nervous system, poriferans are sometimes classified in the subkingdom Parazoa, distinguished from other multicellular animals (Eumetazoa; Hadzi 1963; Steinböck 1963). On the other hand, several biochemical and ultrastructural characters, such as the extracellular matrix (Morris 1993), the acetylcholine/choline esterase system, and flagellated sperm with condensed chromatin and mitochondria have been

[1] Department of Zoology, Graduate School of Science, Kyoto University, Kyoto 606-01, Japan

Progress in Molecular and Subcellular Biology, Vol. 19
W.E.G. Müller (Ed.)

proposed as shared derived characteristics (synapomorphies) that unite all multicellular animals, including sponges, as a monophyletic group (Brusca and Brusca 1990; Schram 1991).

Advances in molecular biology have provided new perspectives for the study of the evolutionary history (Pace et al. 1985; Hills and Moritz 1990). In particular, because of the rapid accumulation of information and the large databases, the small subunit ribosomal RNA or its gene is thought to be ideally suited for phylogenetic studies of distantly related organisms. In addition, various improved phylogenetic methods enable us to demonstrate the pattern of metazoan evolution. Recent analyses based upon almost the entire length of the 18S rRNA sequences suggested monophyly of the Metazoa (Kobayashi et al. 1993; Wainright et al. 1993; Valentine et al. 1996). The inference of pioneer molecular phylogenetic studies based on rDNAs which supported the polyphyly of the Metazoa (Field et al. 1988; Christen et al. 1991) may have been due to a technical problem, particularly too little information, which caused too few suitable, variable sites, and an inappropriate tree-making method. Wainright et al. (1993) have suggested a close relationship between choanoflagellates and metazoans. The cellular similarities between choanoflagellates and poriferan choanocytes are well known. The phylogenetic position of sponges has been shown to be the earliest offshoot in metazoan evolution. Therefore, it is highly likely that the unicellular organisms closest to metazoa are the extant choanoflagellate-like organisms. The 18S rDNA tree basically supports the colonial flagellate hypothesis proposed by Haeckel (1874).

Diploblasts are usually considered to be paraphyletic. Cnidarians and ctenophorans are regarded as being closer to triploblasts, because they contain nerve cells, gap junctions, and an endoderm. However, the relationship between poriferans and placozoans remains somewhat controversial (Ax 1989; Schram 1991; Backeljau et al. 1993). Molecular phylogenetic studies inferred from 18S and 28S rDNA have not completely resolved these issues yet. The phylogenetic status of mesozoans is also controversial. Whether their simple body plan reflects their primitiveness or degeneration due to parasitism is an open question. Molecular phylogenetic analyses based upon 18S rDNA have suggested a close relationship between mesozoans and triploblasts, and therefore supported the latter view (Pawlowski et al. 1996).

As stated above, current molecular phylogenetic analysis of the Metazoa is mainly based on a single gene, 18S rDNA. Several authors have pointed out the instability of phylogenetic trees based on a single gene (Cao et al. 1994; Nikoh et al. 1994). Analyses with specific genes sometimes support the wrong tree. Therefore, molecular data of conserved amino-acid sequences, such as translation elongation factors, RNA polymerases, and ATPases, are necessary to definitively resolve the phylogenetic relationship from sources other than the gene 18S rDNA.

Herein, we discuss early evolution of the metazoa in light of the recent molecular phylogeny inferred from the amino-acid sequences of elongation factor-1α (EF-1α).

2 Inference from EF-1α Sequences

One of the complexes involved in the elongation of amino-acid chains is known as elongation factor-1 (EF-1). EF-1 consists of four subunits named EF-1α, -β, -γ, and δ. EF-1α is involved in aminoacyl-tRNA binding in the step of translation from mRNAs (Hershey 1991). The protein sequences of EF-1α showed apparent similarity and the functional motives, such as the guanine nucleotide-binding site and the phosphoryl-binding site, are highly conserved throughout eukaryotes (data not shown). Because of its universal presence in all living organisms and its slow evolutionary rate, this conserved protein is particularly useful for analyzing the phylogenetic relationships among higher taxa (Hasegawa et al. 1993). Analysis with amino-acid sequences of EF-1α compared with those of nucleotide sequences of 18S rDNA is meaningful from two perspectives. Firstly, the alignment is more reliable. Gaps and insertions in the alignment of nucleotide and deduced amino-acid sequences of EF-1α are rarer than those of 18S rDNA. Taking into account the fundamental role of translation in cells, it is not surprising that many of the factors contributing to translation have allowed very little variation in their amino-acid sequences during evolution. Secondly, amino-acid compositions of highly conserved proteins are thought to be relatively free from the drastic bias of genome G + C content (Hasegawa and Hashimoto 1993). On the other hand, there is a possibility of paralogous comparison using protein sequences. To examine the number of genes coding for EF-1α, we carried out genomic Southern blot hybridization of the poriferan, *Ephydatia fluviatilis*. We detected a single band (data not shown). This suggests that the poriferan gene for EF-1α is present as a single copy. Moreover, cnidarian EF-1α is also encoded by a single copy gene which produces a single transcript (M. Hatta pers. comm.). Both findings suggest that these diploblasts have only one gene for EF-1α and therefore there is little possibility of polymorphism.

3 Monophyly of the Metazoa

The species names, their taxonomic positions, and GenBank accession numbers analyzed in this study are listed in Table 1. Figure 1A shows the phylogenetic tree of eukaryotes based upon a comparison of amino-acid sequences of EF-1α (Kobayashi et al. 1996). Evolutionary distances were calculated by the Dayhoff model (Dayhoff et al. 1978), and phylogeny was inferred by neighbor-joining (Saitou and Nei 1987). This tree supported the monophyly of the Metazoa, including diploblasts and triploblasts, although the bootstrap value for this node was not high enough (42%). Maximum parsimony analysis constructed from the same alignment positions used to infer the tree (Fig. 1A) also supported the monophyly of the Metazoa (data not shown). In both trees, poriferans are included within the metazoan evolution. This is consistent with the hypothesis that sponges had a common origin with other metazoa but

Table 1. List of EF-1α sequences analyzed

Species name	Taxonomic position	Accession numbers
Ephydatia fluviatilis	Porifera (class Demospongea)	D49925
Anemonia erythraea	Cnidaria (class Anthozoa)	D49922
Eugymnanthea japonica	Cnidaria (class Hydrozoa)	D49926
Beroe cucumis	Ctenophora (class Atentaculata)	D49923
Dugesia japonica	Platyhelminthes (class Turbellaria)	D49924
Artemia sp.	Arthropoda	D13338
Bombyx mori	Arthropoda	X03349
Drosophila melanogaster	Arthropoda	X06869
Danio rerio	Chordata	L23807
Xenopus laevis	Chordata	X55324
Homo sapiens	Chordata	X03558
Mucor racemosus	Zygomycota	J02605
Saccharomyces cerevisiae	Ascomycota	X00779
Triticum aestivum	Anthophyta	M90077
Euglena gracilis	Euglenophyta	X16890
Plasmodium falciparum	Apicomplexa	X60488
Tetrahymena pyriformis	Ciliophora	D11083

diverged early in evolutionary history. The tree topologies also suggest the monophyly of diploblasts, although the bootstrap values again are not high (51%). Because of limitations imposed by the length of the sequences, we could not significantly resolve the inner relationship among metazoans. However, branchings of vertebrates and of arthropods were supported by higher bootstrap values (100% in both branchings). The present analysis did not support the monophyly of the triploblasts (Platyhelminthes, Arthropoda, Chordata), although the node was supported by the 18S rDNA sequences (Kobayashi et al. 1993; Wainright et al. 1993). The rate of substitutions in the sequences of triploblast 18S rDNA was relatively higher than those in diploblasts and other eukaryotes. Diploblasts show almost the same substitution rate as other eukaryotes. This higher rate of substitution gives more substitution in the node of triploblasts, which increased the bootstrap value for this node. Phylogenetic analyses were also performed with nucleotide sequences of EF-1α. The second codon position is thought to be the most conservative and is considered to be relatively insensitive to the difference in the overall genome G + C content (Zuckerkandl 1987). Table 2 shows the percentage of genome G + C contents at each codon position. The base compositions at both the first and second codons of EF-1α investigated were not as different as those at the third codon, which varied with each taxon (for example, 22% in *Plasmodium* and 77% in *Drosophila*). Therefore, the base composition of the first and second codons may not affect the phylogenetic analyses. Figure 1B shows the phylogenetic tree from a comparison of the nucleotide sequence at the second codon position only. Evolutionary distances between pairs of DNA sequences were calculated by Jukes-Cantor formula (Jukes and Cantor 1969)

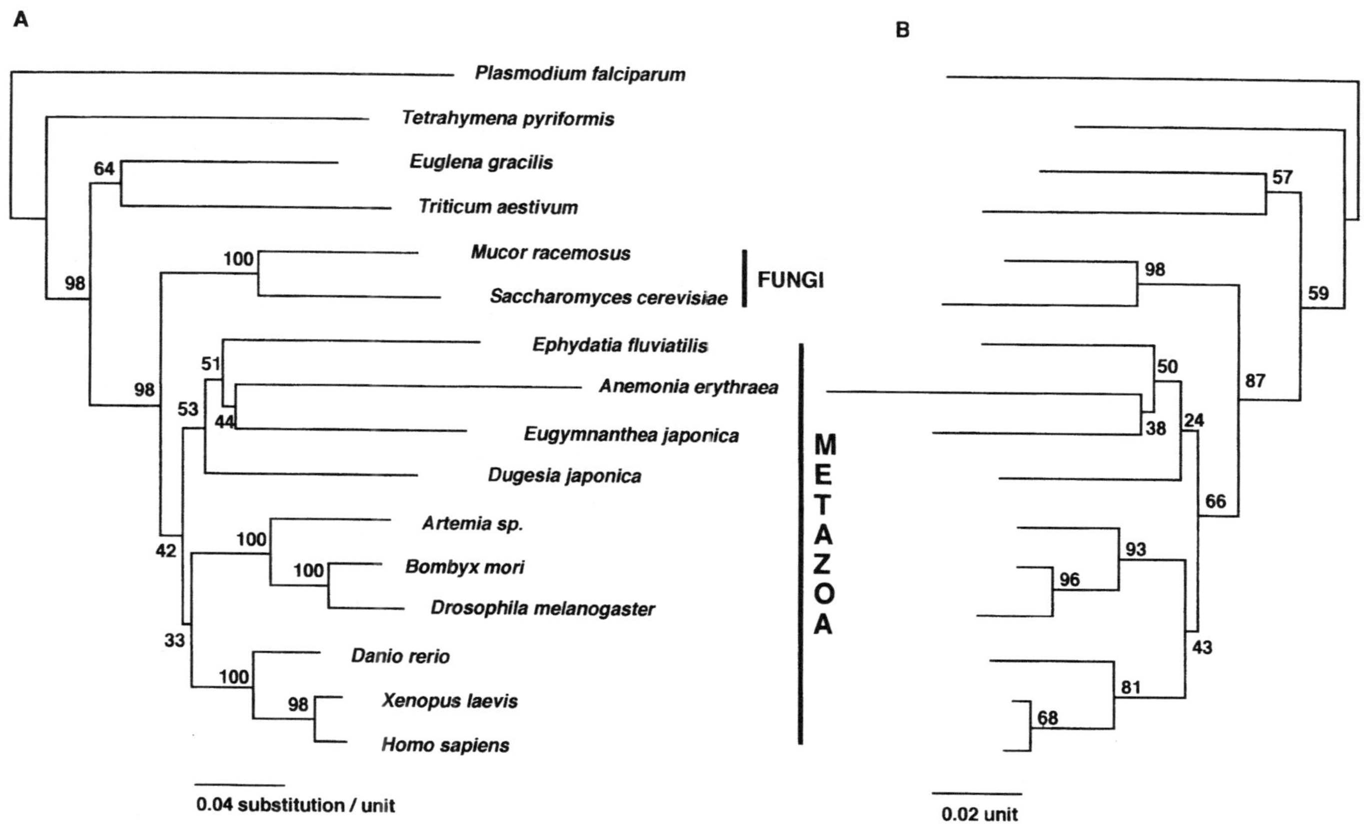

Fig. 1A,B. Molecular phylogenetic trees of eukaryotes. **A** Analysis based on amino-acid sequences of EF-1α. The *scale bar* indicates an evolutionary distance of a 0.04 amino-acid substitution per position in the sequence calculated from Dayhoff PAM matrix model (Dayhoff et al. 1978). **B** Analysis based upon nucleotides at the second positions of codons corresponding to the amino-acid alignments used in **A**. The *scale bar* indicates 0.02 unit of the evolutionary distance calculated from Jukes-Cantor formula (Jukes and Cantor 1969). Each tree was constructed by the neighbor-joining method (Saitou and Nei 1987). The *numbers* at each branch indicate the percentage of times that a node was supported in 100 bootstrap pseudoreplications

Table 2. Percentage of G + C contents

	First codon position	Second codon position	Third codon position
Plasmodium	0.48	0.35	0.22
Tetrahymena	0.52	0.36	0.58
Triticum	0.54	0.39	0.58
Euglena	0.55	0.38	0.67
Mucor	0.55	0.40	0.37
Saccharomyces	0.49	0.39	0.61
Ephydatia	0.54	0.40	0.54
Anemonia	0.52	0.38	0.46
Eugymnanthea	0.52	0.37	0.37
Beroe	0.51	0.37	0.64
Dugesia	0.54	0.39	0.25
Artemia	0.54	0.40	0.42
Bombyx	0.55	0.40	0.49
Drosophila	0.59	0.39	0.77
Danio	0.57	0.40	0.63
Xenopus	0.56	0.39	0.49
Homo	0.56	0.39	0.48

and phylogeny inferred by neighbor-joining analysis. The monophyly of the Metazoa was supported by this tree as well as by the tree based on the first two codon positions (data not shown). The grouping of the Metazoa was supported by relatively high bootstrap values of the neighbor-joining trees (66% at the second codon position and 76% at the first and second codon positions). The grouping of the Metazoa was true when the trees were reconstructed by maximum parsimony analyses and by maximum likelihood (data not shown). Evidence for the monophyletic origin of the Metazoa obtained from a variety of sources and the present study also supports this notion.

4 Phylogenetic Status of the Diploblasts

Figure 2 shows the phylogenetic relationships among metazoan taxa taking the fungi (*S. cerevisiae*) and the protozoa (*T. pyriformis*) as outgroups. Figure 2A was a result of analysis with amino-acid sequences, Fig. 2B with nucleotide sequences at the second positions of codons. As to whether diploblasts are monophyletic or polyphyletic and as to the phylogenetic relationships of diploblasts, the molecular phylogeny inferred from the EF-1α differed from that of 18S rDNA. Within the metazoan subtree, the 18S rDNA tree suggested paraphyly of the diploblastic animals (Kobayashi et al. 1993; Wainright et al. 1993). The Porifera and Ctenophora lineages diverged first, followed by Cnidaria and then triploblastic animals. By contrast, the EF-1α trees (Fig. 2A,B) preferred monophyly (Kobayashi et al. 1996), although high bootstrap values were not obtained (21% in Fig. 2A and 47% in Fig. 2B). Both trees were

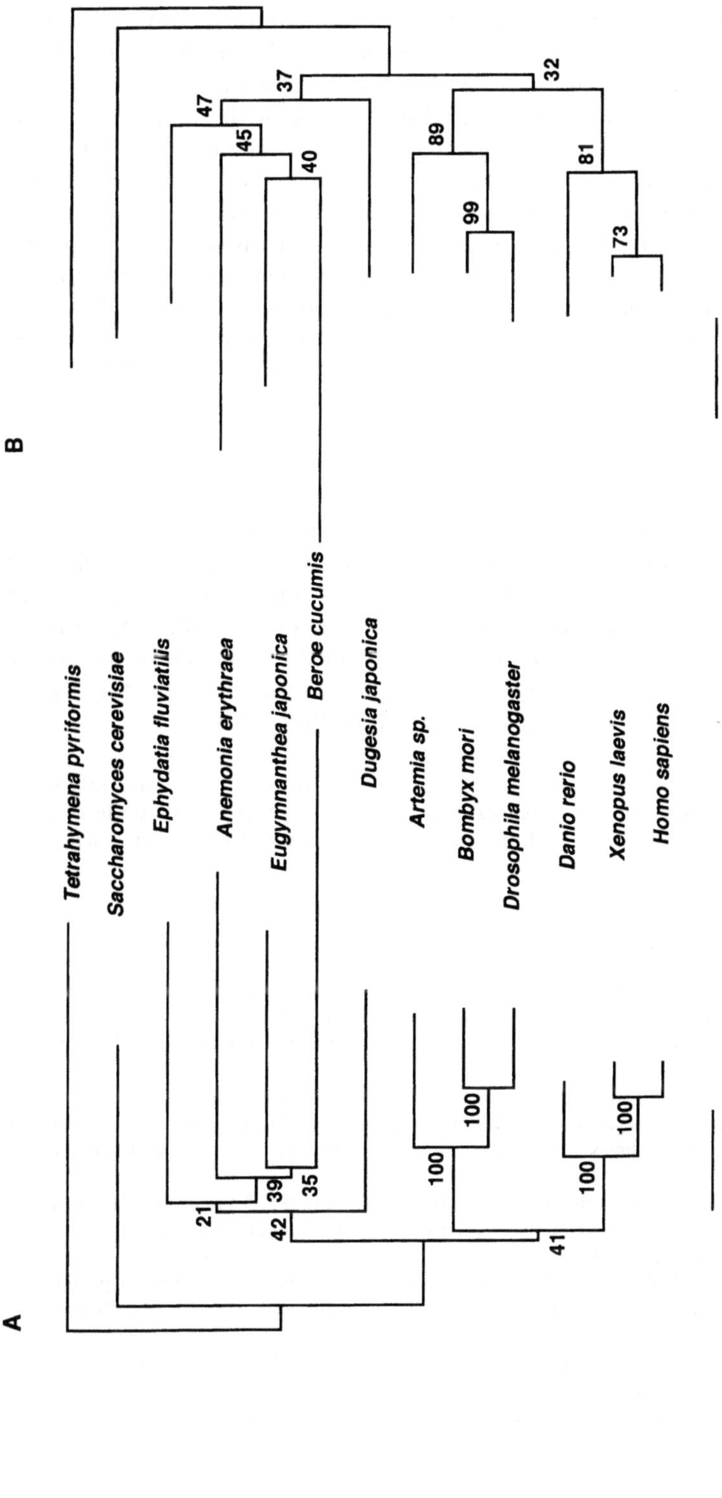

Fig. 2A,B. Phylogenetic relationships of diploblasts with triploblasts. A Analysis based on amino-acid sequences of EF-1α. *Scale bar* 0.04 unit. B Analysis based on nucleotides at the second positions of codons corresponding to the amino-acid alignments used in A. *Scale bar* 0.02 unit

topologically identical, in that a poriferan (*Ephydatia*), two cnidarians (*Eugymnanthea* and *Anemonia*), and a ctenophora (*Beroe*) formed a clade. These topological differences and low bootstrap values obtained from 18S rDNA and EF-1α might be due to rapid divergence of the major metazoan phyla in an extremely short geological period. This is one reason for the unclear branching. Further molecular data that would enable a comprehensive assessment are necessary to resolve this problem. Sequence data of various genes encoding conserved proteins are required. Tracing for the qualitative changes, such as genomic rearrangement and the genetic code of mitochondrial DNA from diploblasts, would be valuable tools for phylogenetic reconstruction. Moreover, an extended survey of the Antp-class homeobox genes in diploblastic animals may shed light on the early evolution of the Metazoa.

5 Conclusion

Molecular phylogenetic analyses based upon EF-1α have given us an insight into the early evolution of the Metazoa. Together with the current molecular and morphological data, the present analyses supported the notion of monophyletic origin of the Metazoa. This is a line of evidence based on the amino-acid sequence that supports monophyly of the Metazoa. However, the phylogenetic relationship of the diploblast groups was not significantly resolved by this study. This will be examined in further studies.

Acknowledgments. M.K. was supported by a JSPS (Japan Society for the Promotion of Science) Predoctoral Fellowship for Young Scientists with Research Grant 4585. This study was also supported in part by a Grant-in-Aid from the Japanese Ministry of Education, Science and Culture to N.S.

References

Ax P (1989) Basic phylogenetic systematization of the Metazoa. In: Fernholm B, Bremer K, Jörnvall H (eds) The hierarchy of life. Elsevier, Amsterdam, pp 229–245

Backeljau T, Winnepenninckx B, Bruyn LD (1993) Cladistic analysis of metazoan relationship: a reappraisal. Cladistics 9: 167–181

Brusca RC, Brusca CJ (1990) Invertebrates. Sinauer, Sunderland, pp 97–277

Cao Y, Adachi J, Yano T, Hasegawa M (1994) Phylogenetic place of guinea pigs: no support of the rodent-polyphyly hypothesis from maximum-likelihood analyses of multiple protein sequences. Mol Biol Evol 11: 593–604

Christen R, Ratto R, Baroin A, Perasso R, Grell KG, Adoutte A (1991) An analysis of the origin of metazoa, using comparisons of partial sequences of the 28S RNA, reveals an early emergence of triploblasts. EMBO J 10: 499–503

Dayhoff MO, Schwartz RM, Orcutt BC (1978) A model of evolutionary change in protein. In: Dayhoff MO (ed) Atlas of protein sequence and structure, vol 5, suppl 3. Nat Biomed Res Foundation, Washington, DC, pp 345–352

Field KG, Olsen GJ, Lane DJ, Giovannoni ST, Gheselin MT, Raff EC, Pace NR, Raff RA (1988) Molecular phylogeny of the animal kingdom. Science 239: 748–753

Hadzi J (1963) The evolution of the Metazoa. Pergamon Press, Oxford

Haeckel E (1874) The gastraea-theory, the phylogenetic classification of the animal kingdom and the homology of the germ-lamellae. Q J Microsc Sci 14: 142–165

Hanson ED (1977) The origin and early evolution of animals. Wesley University Press, Middleton, Connecticut

Harrison FW, Kay NW, Kaye GW (1990) The dermal membrane of *Eunapius fragilis*. In: Rützler K (ed) New perspectives in sponge biology. Smithsonian Institution Press, Washington, DC, pp 223–227

Hasegawa M, Hashimoto T (1993) Ribosomal RNA trees misleading? Nature 361: 23

Hasegawa M, Hashimoto T, Adachi J, Iwabe N, Miyata T (1993) Early branching in the evolution of eucaryotes: ancient divergence of entamoeba that lacks mitochondria revealed by protein sequence data. J Mol Evol 36: 380–388

Hershey, JWB (1991) Translational control in mammalian cells. Annu Rev Biochem 60: 717–755

Hills DM, Moritz C (1990) Molecular systematics. Sinauer, Sunderland

Inglis WG (1985) Evolutionary waves: patterns in the origins of animal phyla. Aust J Zool 33: 153–178

Jukes TH, Cantor CR (1969) Evolution of protein molecules. In: Murno NH (ed) Mammalian protein metabolism, vol 3. Academic Press, New York, pp 21–132

Kobayashi M, Takahashi M, Wada H, Satoh N (1993) Molecular phylogeny inferred from sequences of small subunit ribosomal DNA, supports the monophyly of the Metazoa. Zool Sci 10: 827–833

Kobayashi M, Wada H, Satoh N (1996) Early evolution of the Metazoa and phylogenetic status of diploblasts as inferred from Amino acid sequence of elongation factor-1α. Mol Phylogenet Evol 5: 414–422

Morris PJ (1993) The developmental role of the extracellular matrix suggests a monophyletic origin of the kingdom Animalia. Evolution 47: 152–165

Nielsen C (1995) Animal evolution. Oxford University Press, Oxford

Nikoh N, Hayase N, Iwabe N, Kuma K, Miyata T (1994) Phylogenetic relationship of the kingdoms animals, plantae, and fungi, inferred from 23 different protein species. Mol Biol Evol 11: 762–768

Nursall JR (1962) On the origins of the major groupings of animals. Evolution 16: 118–123

Pace NR, Stahl DA, Lane DJ, Olsen GJ (1985) Analyzing natural microbial populations by rRNA sequences. Am Soc Microbiol News 51: 4–12

Pawlowski J, Montoya-Burgos JI, Fahrni JF, Wüest J, Zaninetti L (1996) Origin of the mesozoa inferred from 18S rRNA gene sequences. Mol Biol Evol 13: 1128–1132

Pedersen KG (1991) Invited review: structure and composition of basement membranes and other basal matrix systems in selected invertebrates. Acta Zool 72: 181–201

Ruppert EE, Barnes RD (1994) Invertebrate zoology, 6th edn. Saunders College, Philadelphia

Saitou N, Nei M (1987) The neighbor-joining method: a new method for reconstructing phylogenetic trees. Mol Biol Evol 4: 406–425

Schram FR (1991) Cladistic analysis of metazoan phyla and the placement of fossil problematica. In: Simonetta EM, Conway-Morris S (eds) The early evolution of metazoa and the significance of problematic taxa. Cambridge University Press, Cambridge, pp 35–46

Steinböck O (1963) Origin and affinities of the lower metazoa. In: Dougherty EC (ed) The lower metazoa. University of California Press, Berkeley, pp 45–54

Valentine JW, Erwin DH, Jablonski D (1996) Developmental evolution of metazoan bodyplans: the fossil evidence. Dev Biol 173: 373–381

Wainright PO, Hikle G, Sogin ML, Stickel SK (1993) Monophyletic origins of the Metazoa: an evolutionary link with fungi. Science 260: 340–342

Willmer P (1990) Invertebrate relationships. Cambridge University Press, Cambridge

Zuckerkandl E (1987) On the molecular evolutionary clock. J Mol Evol 26: 36–46

Subject Index

MIX
Papier aus verantwortungsvollen Quellen
Paper from responsible sources
FSC® C105338
FSC
www.fsc.org

If you have any concerns about our products,
you can contact us on
ProductSafety@springernature.com

In case Publisher is established outside the EU,
the EU authorized representative is:
Springer Nature Customer Service Center GmbH
Europaplatz 3, 69115 Heidelberg, Germany

Printed by Libri Plureos GmbH
in Hamburg, Germany